m/z	Ion	Origin	m/z	Ion
29	HCO$^+$	Aldehydes		
30	CH$_2$NH$_2^+$	Amines	85	(dihydropyran structure)
31	H$_2$C=$\overset{+}{O}$H CH$_3$O$^+$	Alcohols Methyl esters		Tetrahydropyranyl ethers
36	HCl$^{+\bullet}$	Chloro compounds	88	CH$_3$OCO–CH $\overset{\|}{\underset{+NH_2}{}}$	Amino acid esters
43	C$_3$H$_7^+$ CH$_3$CO$^+$	Propyl ion Acetyl groups	91	C$_7$H$_7^+$	Aromatic hydrocarbons with side chains
47	CH$_3$S$^+$	Sulfides	92	C$_7$H$_8^{+\bullet}$	Benzyl compounds with a γ-hydrogen
49	CH$_2$Cl$^+$	Chloro compounds	95	(furyl-CO$^+$ structure)	Furyl-CO-X
55	C$_4$H$_7^+$	Alkyl groups	97	(thiopyran structure)	Alkyl thiophenes
57	C$_4$H$_9^+$ C$_2$H$_5$CO$^+$	Alkyl groups Acylium ion	99	(dioxolane structure)	Etylene ketals of cyclic compounds (steroids)
58	[H$_2$CCOCH$_3$]$^{+\bullet}$ $\overset{H}{}$	Ketones with a γ-hydrogen			
59	[COOCH$_3$]$^+$	Methyl esters	104	C$_8$H$_8$ $^{+\bullet}$	Alkyl aromatics
61	CH$_3$C=O$^+$H $\overset{OH}{}$	Esters of high molecular weight alcohols	105	C$_6$H$_5$CO$^+$ C$_8$H$_9^+$	Benzoyl compounds Aromatic hydrocarbons
70	(pyrrolidine structure)	Pyrrolidines	106	CH$_2$=(ring)=$\overset{+}{N}$H$_2$	Amino benzyl
74	[CH$_2$=C–OCH$_3$]$^{+\bullet}$ $\overset{OH}{}$	Methyl esters with a γ-hydrogen	107	C$_7$H$_7$O$^+$	Phenolic hydrocarbons
77	C$_6$H$_5^+$	Aromatics	117	C$_9$H$_9^+$	Styrenes
78	C$_5$H$_4$N$^+$	Pyridines and alkyl pyrroles	128	HI$^{+\bullet}$	Iodo compounds
80	(pyridine structure)	Pyrroles	130	(quinoline/indole structure)	Indoles
80(82)	HBr $^{+\bullet}$	Bromo compounds	131	(phenyl-CO$^+$ structure)	Cinnamates
81	(pyran structure) (diene chain structure)	Furans Aliphatic chain with two double bonds	149	(phthalate structure)	Dialkyl phthalates (rearrangement)
83	C$_6$H$_{11}^+$ CHCl$_2^+$	Cyclohexanes or hexenes Chloro compounds			

Common Fragment Ions

Reviewers' Comments on the First Edition

"Filled with many useful examples of correct and incorrect usage of important terms, Sparkman's compilation may well become the de facto nomenclature standard for mass spectrometry."

"The Bibliography provides much more than just the expected laundry list of references. It contains multiple sections ... which simplify the search for references on specific topics. Most sections include a brief overview with Sparkman's recommendations of particularly useful selections, and each reference contains the title of the book or article. Internet resources receive special mention on the inside back cover, which provides URL listings of useful mass spectrometry related Web sites."

"O. David Sparkman's *Mass Spectrometry Desk Reference* offers useful information for both the experienced mass spectrometrist and the newcomer to the field. Well organized and easy to read, this text represents a fine resource for those learning the language of mass spectrometry."
–Bill Boggess, University of Notre Dame

"*Mass Spec Desk Reference* is a reasonably priced glossary, tutorial, style guide and bibliography and is a useful addition to the libraries of those who use mass spectrometry in any capacity."

"*Mass Spec Desk Reference* is a valuable tool for both beginners and experts who want to get the most out of the mass spectrometry literature and to communicate effectively with colleagues in all disciplines."
–Kermit K. Murray, Emory University

"The references Sparkman has collected are a valuable service to the mass spectrometric community."

"*Mass Spectrometry Desk Reference* serves two functions—a bibliographic resource and a writing guide. It belongs not just on the desks of mass spectrometrists, but also those of our collaborators and clients."
–Phil Price, Union Carbide Technical Center

"Overall, this is a very useful publication; it is one that everyone who deals with mass spectrometry on a daily basis should definitely have. It is particularly recommended for graduate students who either want to acquaint themselves with the terminology of mass spectrometry and its literature, or as a starting point for individuals planning to specialize in mass spectrometry."
–Trends in Analytical Chemistry

MASS SPECTROMETRY DESK REFERENCE

SECOND EDITION

O. DAVID SPARKMAN

Global View Publishing
Pittsburgh, Pennsylvania

May 2006

Mass Spectrometry Desk Reference

By: O. David Sparkman
E-mail: ods@compuserve.com

Global View Publishing
655 William Pitt Way
Pittsburgh, PA 15238
Telephone: 1-412-828-3191
Fax: 1-412-828-3192

Ordering Information

Individual Sales.   Global View publications are available through most bookstores. They can also be ordered direct from Global View Publishing at the address above or logging onto "The LC/MS Book Store" at www.LCMS.com.

Copyright © 2006 by Global View Publishing
Second Edition

Printed in the United States of America

**Publisher's Cataloging-in-Publication
(*Provided by Quality Books, Inc.*)**

Sparkman, O. David (Orrin David), 1942–
          Mass spectrometry desk reference / O. David
     Sparkman. – 2nd ed.
          p. cm.
          Includes bibliographical references and index.
          LCCN: 00-100995
          ISBN: 0-9660813-9-0

          1. Mass spectrometry—Handbooks, manuals, etc.
     I. Title.

QD96.M3S67 2006   543.0873

This book, as was the first edition, is dedicated to my partner, my best friend, my companion—my wife, Joan;

and

to my partner in mass spectrometry whom many of you have often heard me refer to—our dog, Maggie;

both of whom have aided and endured the first and second editions of this book;

and

to our new addition—Chili—destined to become another famous canine mass spectrometrist.

Table of Contents

Table of Contents

PREFACE, Second Edition

The first edition of *Mass Spectrometry Desk Reference* has now been in print for more than five years. During that time, approximately 4000 copies have been sold; and the book has had a number of favorable reviews. I have received many comments, requests for new definitions, and suggestions as to what might be in this new edition. Also during these five years, a number of new book titles on mass spectrometry have been published as well as a number of titles have been uncovered that were not included in the previous edition's bibliography section. There have been some corrections based on invaluable comments received from readers, such as the misuse of the term *empirical formula* (which relates to the proportion of the elements) when *elemental composition* (the number of atoms of each element present) was actually meant. There has also been a less nonjudicial use of hyphens. The torr was the unit of pressure in the first edition. In the second edition, the unit of pressure is the pascal (Pa). In some cases, there have even been some modifications in the correctness or appropriateness of terms. In the first edition, an effort was made to trace the historical origin of various terms. This effort has been expanded in the second edition. A number of new sections have been added, such as the section on isomer nomenclature. The use of the book has been facilitated by the addition of section headers, which allow the user to easily determine the current section.

Since the first edition appeared, through my personal efforts and the book's impact, articles no longer appear with citations to other journal articles where the cited article's title is not included in the *Journal of Mass Spectrometry* and *Mass Spectrometry Reviews* (John Wiley and Sons), the *Journal of the American Society for Mass Spectrometry* (Elsevier), and the *European Journal of Mass Spectrometry* (IM Publications). There are still mass spectrometry journals and chemistry journals that publish a large number of mass spectrometry articles that have yet to embrace this important standard. Among those who have applauded this effort to include journal article titles in interlibrary loan requests are librarians who categorically stated that obtaining copies of a desired article is much faster and easier when the title is included.

The problems of nonstandardized nomenclature and presentation have become worse in the past few years. In a recent issue of *Analytical Chemistry*, the symbol for mass-to-charge ratio (*m/z*) was used both as an adjective (i.e., *m/z* 1000), as it should be, and as a word modified by a number (i.e., 1000 *m/z*), as it should not be. This same issue had an article that used Th as a synonym for *m/z* without definition. The use of Th (thomson) as a symbol for a proposed *m/z* synonym has not been accepted and is far too esoteric to be used in the same no-defined way as the symbols for liter, meter, gram, pascal, volt, and so forth. Another example of "not taking the effort to make it correct" is found in Volumes 1 and 4 of the *Encyclopedia of Mass Spectrometry*. The Editor of Volume 1 made a concerted effort to make sure that none of the authors used the archaic term "kcal/mole" instead of what is considered to be the modern standard of kJ/mole. The Editor of Volume 4 states in the Preface that this standardization will not be followed, and that the use of both the •+ and the +• conventions will be allowed for this volume. Neither of these two issues should appreciably detract from two excellent works; however, it shows that getting agreement is difficult and is not being addressed by copyeditors who no longer exist.

Another problem that has been increasing in magnitude is the lack of detail regarding the instrument parameters. A section has been added to this edition that details the information required to reproduce an analysis.

Over the past five years, I have taught a fall session of mass spectrometry to a class of doctoral students at the University of the Pacific. Many (if not most) of the students in these classes are what we now call ESL (English as second language). I have become increasingly aware of the problem of proper wording in both spoken lectures and in written assignments. The following is a question from a homework assignment:

> A protonated molecule known not to have any adducts other than the proton, obtained using electrospray, has the empirical formula (correct term) of C_3H_4NO. It is represented by a peak at m/z 2920.00. The next highest peak in the mass spectrum is at m/z 2920.16. Does the ion have an odd or even number of nitrogen atoms? Explain your answer.

One of the ESL students in class read "The next highest peak in the mass spectrum is at m/z 2920.16" to mean the peak that had the next greater intensity. For good clarity, especially for the ESL students, the sentence should have been worded as "The peak at the next highest m/z value is at m/z 2920.16". This may be an extreme example of problems in clarity, but it caused me to recall a lesson I learned early in my professional training.

As a student in the Chemistry Department of what was then North Texas State University (now University of North Texas) in Denton, Texas, I had completed two semesters of undergraduate research. Several intermediates had to be synthesized, and their structures verified. My research advisor had followed the work very closely and knew all the results in detail. When my final paper on the work was submitted, he gave it a quick glance and said that my work had been exemplary. He told me to take a copy of my paper to the Department Chairman, Dr. James L. Carrico, who was actually the listed instructor of the course for which I would receive the grade for my efforts. We spent the next five hours going through every paragraph, sentence, phrase, word, scheme, and figure in the paper. There was not a single square inch that did not have comments written on it. Dr. Carrico continually emphasized that no matter how good the science, if you could not communicate it in such a way that the world could understand it, your work would go unnoticed and unappreciated. I learned as much in those five hours as I had learned in the hundreds of hours spent during the last two semesters doing the syntheses and analyses. The lesson taken from that paper-review also had as profound an effect on me as all the technical aspects of my professional training.

After working on this second edition, realizing the importance of communication, especially with the ESL people of the world, those five hours dwell in my mind and are probably the reason I wrote the first edition and will continue with subsequent editions; and that lesson taught to me by Dr. Carrico is probably why I am so pedantic about using the correct term and not using the incorrect term.

I envision future editions of this book because the field of mass spectrometry continues to evolve and expand. These editions will be greatly enhanced by continued comments from readers, and these comments are greatly encouraged. If you think this book should contain additional definition topics or new terms should be included under existing topics, please let me know. My e-mail address remains unchanged, and I look forward to your comments.

May 2006

PREFACE, First Edition

The analytical technique of mass spectrometry generates mass spectra regardless of the sample introduction technique, the method of ion formation, or the way ions are separated. When a molecule is ionized, a characteristic ion, representing the intact molecule and/or a group of ions of different masses that represent fragments of the ionized molecule, is formed. When these ions are separated, the plot of their relative abundance versus the mass-to-charge ratio (m/z) of each ion constitutes a mass spectrum. Learning to identify a molecule from its mass spectrum is much easier than using any other type of spectral information. The mass spectrum shows the mass of the molecule and the masses of its pieces. Mass spectrometry offers more information about an analyte from fewer samples than any other technique. Mass spectrometry is also the most accurate technique for the determination of mass. The only disadvantage of mass spectrometry compared to other techniques is that, usually, the sample is consumed; however, so little sample is required, it is inconsequential.

Mass spectrometry had its origin in the works of the English physicist, Joseph John Thomson (1856–1940), who won the 1906 Nobel Prize in physics for the discovery of the electron. Thomson's work was further developed by Francis William Aston (1877–1945), another English physicist, a student of Thomson, and also a Nobel Prize winner (chemistry 1922); and by Arthur Jeffrey Dempster (1886–1950), a Canadian-American physicist at the University of Chicago, who worked independently of Thomson and Aston. Aston built the first mass spectrometer based on experimental instrumentation developed by Thomson. Mass spectrometry developed as a distinct field during the period of 1911–1925 based on the efforts of these three outstanding scientists. Today, mass spectrometry has become one of the most widely used analytical techniques for the monitoring of respiratory gases to the surface analysis of construction materials to the sequencing of complex DNA molecules. Advances in techniques of ion formation and separation have stripped away the volatility and limited mass requirements of the analyte that were present in the formative years through the 1970s. Now that the technology has advanced to the point that mass spectrometrists "...can make elephants fly", there is no applications area of analytical chemistry that remains untouched by mass spectrometry.

A large amount of mass spectral data of small molecules comes about as a result of gas chromatography/mass spectrometry (GC/MS). GC/MS data are usually taken under electron ionization (EI)[1] conditions at about 70 eV. Some EI mass spectral data result from introduction of pure samples via a direct insertion probe, and some by means of particle beam liquid chromatography/mass spectrometry. Most all of the spectral libraries and compilations are of EI spectra taken at 70 eV. A fundamental requirement of EI mass spectrometry is that it must be possible to put the analyte molecule in the gas phase under reduced pressure conditions. This requirement is not true of analytes that

---

[1] EI can refer to both electron ionization and electron impact ionization. The term "electron impact" was replaced with electron ionization after the abbreviation "CI" was adopted for chemical ionization. Electron impact is considered to be archaic.

are analyzed using one of the desorption/ionization techniques, which are often used for the analysis of macromolecules [e.g., liquid chromatography electrospray (ES), matrix-assisted laser desorption/ionization (MALDI) mass spectrometry, or fast atom bombardment (FAB)]. However, regardless of the method of ion formation or the size of the ionized molecule, fragments produced will always result from the loss of a logical grouping of atoms.

Because mass spectral data obtained by sample introduction through a chromatograph (gas or liquid) are a collection of mass spectra taken during the chromatographic separation of mixtures, they can be displayed as mass chromatograms (a plot of scan number versus the intensity of a single mass-to-charge ratio) or reconstructed total-ion-current chromatograms (RTICC),[2] both of which can be used for precise quantitation. This added dimension is a decisive advantage of mass spectrometry over other techniques and can be used to verify the validity of data and remove analyte contaminants. In the absence of a chromatograph as the sample-introduction device, it is especially important to be able to distinguish interpretable mass spectral peaks from background and decomposition peaks.

Spectra obtained using the so-called "soft ionization" techniques of desorption and chemical ionization (DI and CI) can result in protonated and other adduct molecules that have a charge. The fragmentation of these ions is interpreted using the same general principles as used in the interpretation of EI spectra. Since these soft ionization techniques involve a less-harsh treatment of the analyte, a molecular ion (or representation of it) is much more likely to be present in the spectrum. Large nonvolatile molecules can be analyzed with mass spectrometry. The development of techniques such as ES (and its variants) for liquid chromatography/mass spectrometry (LC/MS) has greatly aided the analysis of mixtures of nonvolatile analytes in complex biological and environmental matrices. New ionization techniques such as MALDI, coupled with mass spectrometers not limited by a maximum m/z value (the time-of-flight mass spectrometer), have made the analysis of carbohydrates, nucleic acids, peptides, and proteins a reality.

All of this means more people in more diverse disciplines will be using mass spectral data. Due to the lack of training (because mass spectrometry was not pertinent to their field of study at the time of their initial education), these people need tools such as provided by this book to help them in getting the most from the data and technique. As you begin or continue your study of mass spectrometry, try to develop your own tools so that your skills become second nature—like riding a bicycle.

January 2000

---

[2] Many data systems call this TIC. TIC is used to reduce the amount of space required for the label.

ACKNOWLEDGMENTS, Second Edition

During the years leading up to the second edition, significant enhancements have been made primarily due to comments, suggestions, and questions from readers of the first edition. Their support and encouragement are as important to this second edition as were the peer reviewers to the first edition.

In addition, I would like to express my appreciation to my colleagues with whom I teach the American Chemical Society short courses for their continual comments, inputs, and inspiration—J. Throck Watson and Frederick E. Klink; as well as Patrick R. Jones, Chair Emeritus, Department of Chemistry, University of the Pacific, for his support and help over the past five years; Harold G. Walsh, Director of the ACS Short Course Program, for his support of the courses I have developed and teach, which have influenced this book; and the organic chemistry staff at the University of the Pacific for their detailed review of the Isomer Nomenclature section. And last but by far not least, as was the case with the first edition, I owe my greatest appreciation for this book to my wife, Joan A. Sparkman, who has been the copyeditor for each and every one of the hundreds of revised and expanded drafts.

I also must express my appreciation to Ross and Mary Ellen Willoughby at Global View Publishing. Ross is responsible for the design of the cover for both editions. Mary Ellen has the responsibility of keeping all the orders straight and shipping the books to you. Ross and his partner, Ed Sheehan, have also been very helpful with the various revisions.

O. David Sparkman
Department of Chemistry
University of the Pacific
Stockton, California
May 2006

ACKNOWLEDGMENTS, First Edition

The **Correct and Incorrect Terms** section of this book has been in the preparation stage for almost 10 years. While the GC/MS trainer for the Varian CSB worldwide sales organization, it became obvious that it was necessary to originate such a document as a way to prevent people from (1) using "self-CI" to describe two separate phenomena that resulted in spectral anomalies (space-charge effects resulting from ion overload, and ion/molecule reactions caused primarily by odd-electron fragment ions protonating analyte molecules) in the quadrupole ion trap mass spectrometer, and (2) using "amu" as a synonym for mass-to-charge ratio or as a symbol for atomic mass unit. Over the years, the document grew from its original four 8½ × 11 pages to somewhat less than its present size based on issues raised in the various American Chemical Society mass spectrometry short courses that I teach. In conversation (both verbally and electronically), reference would be made to this document; then, as a result, requests for copies would be made. Some of these requesters encouraged me to publish this document because of their perceived significance of the material.

In March of 1998, Dominic Desiderio, one of the co-editors of *Mass Spectrometry Reviews*, contacted me and said he had heard of the document and wanted to know if it could be submitted for possible publication. In April of that same year, my first draft was submitted. Because of the potential impact and some of my contradictions to what had been published by the American Society for Mass Spectrometry (ASMS) and the International Union of Pure and Applied Chemistry (IUPAC), Dominic elected to send the manuscript to more reviewers than normal. Based on personal comments that were received from some of those selected to review the draft, it was obvious that many, if not all, of the editors of major mass spectrometry journals had been selected as reviewers. All of the reviews were very thorough and had many helpful comments and suggestions as well as a few "who-do-you-think-you-are" statements that were made about my disagreements with the ASMS and IUPAC publications.

On April 2, 1999, a revised manuscript was submitted to *Mass Spectrometry Reviews* that addressed many of the issues raised by the reviewers and corrected a number of technical inaccuracies pointed out in the reviews. On May 1, 1999, a letter was sent to me that was jointly signed by the co-editors of *Mass Spectrometry Reviews* (Dominic M. Desiderio and Nico M. M. Nibbering), stating their concurrence with the "...consistent consensus among all of those reviewers that *Mass Spectrometry Reviews* might not be the appropriate publication in which to publish an article on correct and incorrect terms"—an opinion with which I cannot disagree.

The quality of the information, some additions and deletions, and the presentation itself is largely due to the comments of those reviewers who did a complete and thorough job in their reviews. Not one of the returned reviews was accomplished in a short period or without considerable thought. Although I still do not have a consensus with all of the reviewers on all of the terms and their definitions, the article "Correct and Incorrect Terms for Mass Spectrometry" has had the benefit of the peer-review process whose purpose is not only to evaluate the technical content but also to provide the author with the insight and wisdom of the reviewer, thereby resulting in a superior product. The co-editors of *Mass Spectrometry Reviews* and their selected reviewers deserve my deepest gratitude. This acknowledgment is in no way meant to imply that any of these people has expressed an endorsement of this particular work.

In addition, I would like to express my appreciation to my two colleagues with whom I teach the American Chemical Society short courses for their continual comments, inputs, and inspiration—J. Throck Watson and Frederick E. Klink. And last but by far not least, I owe my greatest appreciation for this book to my wife, Joan A. Sparkman, who has been the copyeditor for each and every one of the hundreds of revised and expanded drafts.

O. David Sparkman

CONVENTIONS USED IN THIS BOOK

Some of the terms defined as **Correct** or **Incorrect** in this book are exceptions to the definitions found in the *Current IUPAC Recommendations* (Todd 1995) and the *ASMS Guidelines* (Price 1991). These terms are noted with an asterisk (*). This book is not meant to supersede compilations of official terms, but should be used as a clarification tool. It is important to remember that the aim is to communicate with a minimal amount of confusion.

In some cases where a term is listed as **Incorrect**, the text immediately following the term describes the incorrect usage. Text following this statement of incorrect usage will contain a correct definition and/or usage of the term. There may be circumstances where this term should be used.

Example:

Incorrect: **API** – when used as an abbreviation for atmospheric pressure chemical ionization (APCI). APCI was originally called atmospheric pressure ionization and abbreviated as API. This use is archaic and causes confusion.

In other cases, a term listed as **Incorrect** should not be used regardless of the circumstance.

Example:

Incorrect: **protonated molecular ion** – when used to describe a molecule that has been protonated. This term implies that the molecular ion (a positive-charge species) has reacted with a proton (another positive-charge species). For this reaction to happen, the two species would have to come in contact with one another. This event is unlikely because like charges repel one another.

References for the **Terms** section of this book are listed in alphabetical order.

Listings in each segment of the **Bibliography** section of this book are in chronological order.

CORRECT AND INCORRECT TERMS
FOR
MASS SPECTROMETRY

INTRODUCTION

The audience for mass spectrometry information is wider than it has ever been and continues to grow. Many members of this audience do not have English as their first language; many do not have extensive chemistry or scientific backgrounds (i.e., attorneys, physicians, engineers, etc.). Therefore, in working with various forms of mass spectrometry, it is important that the correct scientific terminology is used to describe the hardware, data, and technique. Unfortunately, the field of mass spectrometry does not follow a strict adherence to the use of SI units (see *m/z*). Even the guidelines for terminology set forth by the International Union of Pure and Applied Chemistry (IUPAC) and the American Society for Mass Spectrometry (ASMS) contain esoteric neologisms that have become the accepted standards, contradictions, and typographical errors. This document is a guideline of **correct** and **incorrect** terms used in print and in oral presentations on the techniques and instrumentation of mass spectrometry and hyphenated mass spectral techniques such as gas or liquid chromatography/mass spectrometry (GC/MS or LC/MS). The primary purpose of this document is to draw attention to the need for standardized nomenclature so that all cultures (social and technical) can assimilate the scientific information that is being presented in English and to point out how often standards published by committees are at best confusing and, as the case with the IUPAC definition of molecular mass, are in complete contradiction to all previously accepted definitions (see **molecular mass**). Another purpose of this presentation is to collect terms that are often used with mass spectrometry (but that are not specific to such data or techniques) and define them along with mass spectrometry terms so that newcomers to the field can have a single reference.

Over the past several decades, there have been a number of compilations of terms for use with mass spectral techniques (Beynon 1977, 1981; Karasek, Clement 1988; Price 1991; Todd 1995; de Hoffmann 1996; McLafferty, Tureček 1993; Micromass 2000). Since the publication of the first edition of this book, several mass spectral bibliographies have appeared on the Internet from Shimadzu, Cambridge Health Care, *spectroscopyNOW*, and a fledgling effort from an IUPAC group in the Analytical Chemistry Division, to name just a few. A search of "mass spectrometry terms" will lead to these bibliographies. Developing an accurate list of correct terms is difficult from two standpoints: (1) the ever-changing and advancing technology that requires new terms to communicate results and ideas; (2) the problem that many of the terms in mass spectrometry are not really being defined, and that authors have to be put in the position that they "know the definition". A good example of mass spectrometry nomenclature confusion is the various definitions of nominal mass and monoisotopic mass. Three different sources define the nominal mass of an element in three different ways: (1) as the integer mass of each of its isotopes (Biemann 1967); (2) as the integer mass of the lowest mass naturally occurring stable isotope of an element (Watson 1997); (3) as the integer mass of the most abundant naturally occurring stable isotope—the correct definition (Bursey 1971). Rather than defining monoisotopic mass as being based on the exact mass of the most abundant naturally occurring stable isotope of an element, another source defines this term as an integer mass value (de Hoffman 1996).

The two references most often cited in this presentation of mass spectrometry terms are the *Current IUPAC Recommendations* (Todd 1995) and the *ASMS Guidelines* (Price 1991). The terms listed as "incorrect" are some of the more flagrant examples of problems in terminology. Some terms are listed as "incorrect" in order to eliminate the use of two different terms to describe the same thing, whereas other terms have been specifically designated as "no longer recommended" by the *Current IUPAC Recommendations*. Some of the various sources on terminology are not always in agreement with each other; and because this author, along with others, is not in complete agreement with the cited references that are presented here, a personal judgment will have to be made in case of a conflict. Some of the material contained in this document is clearly the preference of the author (e.g., restricting the use of magnetic sector to single-focusing instruments that use a magnetic field; the use of collisionally activated dissociation (CAD) in preference to collision-induced dissociation (CID); the incorrectness of the use of parent and daughter to describe ions, etc.). To include definitions such as those for atom, ion, (etc.) and terms used in organic chemistry such as heterolytic and homolytic cleavages may appear patronizing but is not the intent. Definitions of these terms are necessary for a single source of information. This glossary is not intended to be a tutorial in mass spectrometry; its intent is to be a collection of terms encountered in mass spectrometry. In addition, where possible, the origin of the term and its originator are cited. For ease of use, terms in **boldfaced** type are indexed.

Although this document is written as a guide for authors, it is intended more as a tool for people who deal with mass spectrometry and its data and only have a minimal knowledge of the field.

This glossary omits definitions that pertain to data systems and vacuum systems. For information relating to vacuum systems, refer to the American Vacuum Institute, the Beynon report of the ASMS Nomenclature Workshop (Beynon 1981), or the *Current IUPAC Recommendations*. The Beynon report is also a good reference for the definitions of various terms that are associated with MS/MS in double-focusing mass spectrometers. Data system terminology is very specific to the products that are produced by the various mass spectrometer manufacturers and the computer industry. The data system terminology that appears in the *Current IUPAC Recommendations* and the *ASMS Guidelines* is dated because it revolves around the minicomputer. Care should be taken in the use of terms that relate to data that are the creation of instrument manufacturers. A good source for references for data system hardware and operating systems is the manufacturer of the computer. In the more recent years (since 1990), the problem of custom computers and operating systems for mass spectrometry has been relieved by the existence and proliferation of modern microcomputers. The one exception to data system terminology in this book is the area of computerized mass spectral library searches. Definitions used by various manufacturers are included in an effort to clarify some of the apparent confusions.

There are terms that are more relevant to mass spectral instrumentation research and ion physics that appear in the *Current IUPAC Recommendations* and the *ASMS Guidelines* but do not appear in this document. Therefore, both of these publications are important in the definition of mass spectral terms. This document is more oriented toward the applications of mass spectrometry.

One other area that is only peripherally addressed here is the use of hieroglyphs in mass spectrometry communications. Lehmann (Lehmann 1997) has proposed a set of pictograms that could be added to the graphic display of mass spectral data and would allow a viewer to have an instant awareness of how the data were obtained and under what conditions. Although this convention has yet to be widely embraced, it offers a great deal of potential as do the graphics proposed for MS/MS acquisition modes (de Hoffmann 1996).

DATA

Correct: **background** – a description of a signal observed in a mass spectrum that comes from the presence of chemical materials other than the analyte. This background could be due to GC column bleed (usually cyclic alkyl-siloxanes, produced by the thermal decomposition of the column's stationary phase, which have molecular mass <500 that are characterized by mass spectral peaks of decreasing intensity at *m/z* 207, 281, 355, and 439). Background signal can also come from noise in the electronics.

Correct: **background subtraction** – the process of subtracting the absolute intensity of each mass spectral peak in a background spectrum from the absolute intensities of each mass spectral peak in a sample spectrum. The background spectrum and the sample spectrum can be a single spectrum or an average of a series of spectra. Although this subtraction can result in negative intensity values for peaks, the resulting **background-subtracted spectrum** displayed will not have negative peaks. Background subtraction can result in a spectrum that is easier to interpret because it is free from peaks that represent contaminants.

When a reconstructed total-ion-current chromatographic peak represents more than one analyte, a spectrum representing the preponderance of a single analyte can be subtracted from the spectrum that represents the preponderance of a second analyte (not necessarily the spectrum representing the apex of the chromatographic peak) to obtain a pure spectrum of the second analyte. The reverse subtraction can then be carried out to obtain a pure spectrum of the second analyte. Care must be taken in that a significant peak in one of the mass spectra is not eliminated by the process when the analytes have ions of common *m/z* values.

Correct: **chemometrics** – according to Barry M. Wise (Eigenvector Research, Inc.), "...the chemical discipline that uses mathematical and statistical methods to relate *measurements* made on *chemical* system to the state of system and [to] design or select optimal *measurement* producers and experiments" (Wise 2001).

Correct: **centroid** – a description of the center-of-mass of a mass spectral peak. The centroid of a displayed mass spectral peak representing ions of a common elemental and structural composition but varying isotopic compositions (e.g., peaks observed for multiple-charge ions in mass spectrometers that have constant resolution throughout the *m/z* scale) is the ion's *average mass* (M_r). The centroid of an individual mass spectral peak after the mass spectrometer's detector's analog signal has been digitized through the instrument's analog-to-digital converter is often stored as the recorded *m/z* value for that peak. The centroid of such a peak is assigned an *m/z* value based on the instrument's calibration of the *m/z* scale.

Incorrect: **deconvoluted peak** – this is an excellent example of where the word *peak* should not be used without the descriptive adjectives *mass spectral* or *chromatographic*. A **deconvoluted mass spectral peak** (**correct**) refers to a peak whose mass-to-charge ratio has been converted to mass by multiplying the *m/z* value by the number of charges. A **deconvoluted chromatographic peak** (**correct**) refers to a reconstructed total-ion-current peak that has had individual mass spectra of multiple components extracted free from mass spectral peaks of one another. A **deconvoluted mass spectrum** (**correct**) is one that has had mass spectral peaks and/or intensities due to a contaminating component removed.

Correct: **dark matter** – a mass spectral experiment results in charged particles (ions) and neutral particles (that are produced by the fragmentation of the initially formed ion). These neutral particles (radicals and molecules) are referred to as the **neutral losses** from the originally formed ion. These neutral losses are the dark matter of the mass spectrum.

Incorrect: **empirical formula** – when used to describe a molecular, ionic, or radical formula or an elemental composition. The molecular formula for, or the elemental composition of, ethylene glycol is $C_2H_6O_2$. The empirical formula for ethylene glycol is CH_3O. The empirical formula for benzene (C_6H_6) and acetylene (HC≡CH) is CH. The empirical formula relates to the proportion of the elements present, *not* the number of atoms of the elements present.

Correct: **elemental composition** – describes a molecule, ion, or radical with respect to the kinds and numbers of atoms from which it is composed. The elemental composition of a molecule is also known as the *molecular formula*.

Correct: **fragmentation pattern** – the result of the decomposition of a precursor ion to produce a series of product ions with specific abundances. This pattern is displayed as a mass spectrum.

Correct: **formula** – describes matter analyzed by mass spectrometry; a series of elemental symbols and numbers representing molecules, ions, or radicals. There are three types of formulas: *empirical*, *molecular* (*ionic* or *radical*), and *structural*. The **empirical formula** is the simplest whole number ratio of the different atoms that constitute the molecule, ion, or radical (the proportion of the elements, *not* the number of elements). Acetic acid's empirical formula is CH_2O. The **molecular** (**ionic** or **radical**) **formula** is the listing of the number of atoms of each element that comprises the species. The molecular formula for acetic acid is $C_2H_4O_2$. The molecular (ionic or radical) formula is the same as the *elemental composition*. The **structural formula** is a representation of the arrangement of all the specified atoms of the various elements in the molecule, ion, or radical and their bonds. The structural formula for acetic acid is CH_3COOH or CH_3CO_2H. The convention of writing the molecular formula or the elemental composition with C first, H second, followed by the remaining elements (Br, Cl, F, I, N, O, P, S, and Si) in alphabetical order is called the **Hill formula** order after Edwin A. Hill (Hill 1900).

Structural formulas are sometimes presented as a series of different symbols.

$$H-\overset{\overset{\displaystyle H}{|}}{\underset{\underset{\displaystyle H}{|}}{C}}-\overset{\overset{\displaystyle O}{\|}}{C}-OH \qquad H_3C-\overset{\overset{\displaystyle O}{\|}}{C}-OH \qquad CH_3COOH \text{ or } CH_3\overset{\overset{\displaystyle O}{\|}}{C}OH$$

Dash formula   Condensed formula   Bond-line formula

Acetic acid is represented as three different types of structural formula. In the case of the **dash formula** and the **condensed formula**, the symbol for all the atoms is written. In the **bond-line formula**, the symbol for carbon and any hydrogen atoms necessary to satisfy carbon's valence of 4 are not written. These atoms are implied. All other atoms are written, including H atoms associated with functional groups created with heteroatoms (non-carbon–hydrogen atoms; i.e., O, N, S, P, etc.).

Another symbolism used in the representation of a molecule, ion, or radical is called the **Markush structure**. The Markush structure is one where a substituent is present on the precursor molecule. The position and/or the nature of (element, chemical structure, functional group, class of chemical structure [i.e., alkyl or aryl]), a class of functional groups (i.e., esters), and so forth only has to be specified in a detailed list. The value of such a structure is to allow patent protection for compounds related to those in a specified invention without requiring that each and every possible compound be synthesized and tested.

The name is derived from Dr. Eugene A. Markush, who in 1923 filed a patent application for a method of preparing pyrazoline dyes for use on wool or silk. The U.S. Patent Office challenged the claim as being too unspecified. The U.S. Commission of Patents ruled on the proprietary of such claims when Markush appealed the original ruling. The patent was issued in August of 1924 (1,506,316). A good example of a Markush structure is the regioisomers of xylene.

Incorrect: **line** – when used to describe a mass spectral peak. This term has recently become popular in some circles but is incorrect and does not refer to the actual digital plots that are often observed in a mass spectrum. Spectral lines are seen on a photographic plate in emission spectrography used for the elemental analysis of metals or in mass spectrographs, *not* in mass spectrometry. Another particularly irksome and incorrect term for the description of the graphic representation of a digitized mass spectral peak is *stick* or *stick plot*.

Correct: **ions** are found inside mass spectrometers; *peaks* are found on paper or on the data record. The word *intensity* is used with respect to the height of a peak (**peak intensity**) or to the strength of an ion beam. The word *abundance* is used to describe the number of ions in the mass spectrometer (**ion abundance**). Peaks are often displayed as straight lines in digitized mass spectral data. Mass spectral peaks can also be displayed as profiles with width as well as height. In GC/MS or LC/MS, care must be taken to clearly differentiate between *mass spectral peaks* and *chromatographic peaks*. The most intense peak in a displayed mass spectrum is the **base peak**. When the term (*mass spectral*) *peak intensity* is used, it usually means relative intensity of the specified peak to the intensity of a specified base peak. When the term *percent ion abundance* (such as the percent ion abundance for a molecular ion) is used, it can refer to the peak intensity of the molecular ion peak divided by the sum of the intensities of all the peaks in the mass spectrum.

Correct: **isotope peak** – describes a peak in a mass spectrum where one or more of the atoms of the various elements in the ion represented by the peak is an isotope other than the element's most abundant. The ratio of the intensity of an isotope peak to the intensity of a monoisotopic peak or an isotope peak at a lesser *m/z* value can be used in determinations of elemental compositions.

Correct: **extracted ion current** (**EIC**) **chromatogram** – describes the plot of the intensity at a single mass-to-charge ratio (*m/z*) value, a sum of discontinuous *m/z* values, or a range of *m/z* values that is a subset of the total range of each acquired spectrum versus the spectrum number from a chromatographic separation. In the event that the sample is introduced into the mass spectrometer by some process other than a chromatographic one, the word *chromatogram* should be replaced with *profile* (e.g., an *evaporation profile* when the sample is evaporated [or sublimed] from a solid probe). In the case of capillary electrophoresis (CE), extracted ion current chromatogram is replaced with extracted ion current **pherogram**. Extracted ion current chromatograms are not the same as SIM chromatograms.

Incorrect: **mass chromatogram** – this term was considered "correct" in the first edition of this book; however, the term *mass chromatogram* no longer is appropriate because much of the current mass spectral data represents multiple-charge ions. The *extracted ion current* term does not exclude individual *m/z* data, multiple-charge ions, a range of *m/z* values, or the sum of a series of noncontiguous *m/z* values, whereas the *mass chromatogram* term implicitly does. Mass chromatogram was first used to describe this reconstructed chromatographic technique by Ronald A. Hites (Hites 1970). The author of this book will continue to use the term *mass chromatogram* when referring to GC/MS data and other appropriate data in other writings.

Incorrect: **reconstructed ion chromatogram** (**RIC**) or **reconstructed extracted-ion-current** (**REIC**) **chromatogram** (**EICC** or **REICC**) – when used as synonyms for mass chromatograms. All four of these terms have been used and are self-explanatory; however, confusion is best avoided when there is a single word or phrase used to describe a specific type of data, and when the display of these data is called the *mass chromatogram*.

Correct: **mass spectrum** (singular) and **mass spectra** (plural) – the digital (each mass spectral peak represented by a single line) or analog (**profile data**) plot of the intensities observed at each acquired mass-to-charge ratio (see **Figures 1A** and **1B**). These data are often normalized relative to the most intense peak—the **base peak**. The mass spectrum can be displayed in a normalized or non-normalized table of *m/z* values and intensities. A mass spectrum can also be the presentation of data where the *m/z* values have been converted to mass by multiplying the *m/z* value by the value for *z*. The term *mass spectrum* was first used by Francis William Aston (English Nobel laureate in chemistry, 1922, 1877–1945; one of the founding fathers of mass spectrometry) in 1920 (Kiser 1965).

Correct: **molecular ion peak** – any peak in the mass spectrum that represents an ion with the same elemental composition and connectivity of atoms as the intact molecule. A single-charge molecular ion peak is the highest *m/z* peak that is neither an isotope peak nor due to background that represents an odd-electron ion and has the same nominal mass, elemental composition, and connectivity of atoms as the intact molecule. The highest *m/z* value peak in the mass spectrum does not necessarily repesent a single-charge molecular ion. The highest *m/z* value peak of significance may represent a protonated molecule, an adduct ion, or a fragment of the molecular ion. The sign of the charge of a molecular ion can be positive or negative. Molecular ions can have multiple charges. There is usually one radical site for every charged site.

Correct: **monoisotopic peak** (or **ion**) – the peak that represents an ion that contains only the most abundant isotopes of each of the elements comprising the ion. The monoisotopic peak (or ion) can be either the *nominal* mass peak or the *monoisotopic* mass peak (see **Figure 11** on page 65).

Correct: **monoisotopic mass spectrum** – a mass spectrum with peaks that represent the principal isotopes of the atoms that compose each ion. This presentation is usually a mass spectrum containing only nominal *m/z* value peaks representing each ion (molecular and fragment).

Incorrect: **MS/MS spectrum** or **MS^n spectrum** – when used to describe the mass spectrum that results from the controlled dissociation of a precursor ion or a series of first, second, (etc.) generation product ions (various mass spectrometry/mass spectrometry techniques). An "MS/MS mass spectrum of *n*-butylbenzene" does not convey the same message as the "product-ion mass spectrum of the *n*-butylbenzene *m/z* 134 ion". Product-ion mass spectra often do not exhibit isotope peaks because a monoisotopic precursor ion of a single *m/z* value was selected.

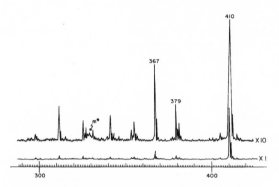

Figure 1A. Profile mass spectral data.

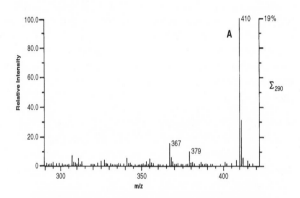

Figure 1B. Digitized mass spectral data.
(Watson, JT *Introduction to Mass Spectromety*, 3rd ed.;
Lippincott-Raven: Philadelphia-New York, 1997; p 75)

Correct: **profile mass spectral data** – describes a mass spectrum where each mass spectral peak is displayed with a height and a width. The height at any point on the peak is a representation of the ion abundance. Each point along the width of the peak represents an ion of the same *m/z* value as that at the horizontal center of the peak but with a different energy and/or ions of different *m/z* values that are not resolved (**Figure 1A**).

Incorrect: **scan** – when used to refer to a mass spectrum. Scan is not correct for mass spectra that are acquired with a time-of-flight mass spectrometer (TOF-MS) or from a Fourier transform ion cyclotron resonance mass spectrometer (FTICR-MS). The magnetic-sector mass spectrometer acquires data by scanning [ramping] the magnetic field strength. The quadrupole ion trap mass spectrometer acquires data by scanning the amplitude of a fixed-frequency radio frequency (RF) voltage. The transmission quadrupole mass spectrometer acquires data by holding the ratio of an RF amplitude to that of a direct current (DC) voltage constant, while increasing both. Although "scan" is correct for these latter three instruments, there is nothing "scanned" in a TOF-MS or an FTICR-MS; therefore, the word *scan* can lead to confusion in TOF mass spectrometry (TOFMS) or FTICR mass spectrometry (FTICRMS) and should not be used. Two terms that often include the word *scan* are **scan rate** and **scan range**. The correct terms are **spectral acquisition rate** and **spectral acquisition range**, respectively. The former meaning the rate at which spectra are acquired or the spectral acquisition time; and the latter, the contiguous *m/z* range over which each spectrum was acquired.

Correct: **reconstructed total-ion-current (RTIC) chromatogram (RTICC)** – defines a chromatographic plot prepared (reconstructed) from a consecutively recorded array of the sum of intensities of all the peaks in each spectrum versus the spectrum number (which is a function of a time domain). In the case of samples introduced into the mass spectrometer by some means other than a chromatographic technique, the plot would be a *reconstructed total-ion-current chronogram*. The important word in this phrase is *reconstructed*.

Note: **total ion current (TIC) chromatogram (TICC)** – a chromatographic output obtained when the total ion current is monitored in "real time" in the ion source. This type of output was used before the development of data systems to determine when to acquire a mass spectrum. The total ion current in the ion source is not monitored in most modern mass spectrometers, and the term was usually associated with magnetic-sector instruments of older vintages. Some manufacturers of GC/MS instrumentation incorrectly use total ion chromatogram (TIC: abbreviation is in conflict with that used for total ion current) to refer to the reconstructed total-ion-current chromatogram (RTICC).

Correct: **SIM chromatogram** or **selected-ion-monitoring chromatogram** – a chromatographic display that results from a selected-ion-monitoring (SIM) analysis. **Note:** J. Throck Watson (Watson 1997) refers to this data display as an **ion current profile**; however, with the exception of probe (in GC/MS) or direct infusion or flow injection (in LC/MS) techniques or in the case of MALDI, the data are indeed chromatographic (elution as a function of time); and the use of ion current chromatograms has the possibility of creating confusion with the technique of ion chromatography. For further clarification between a mass chromatogram and an SIM chromatogram, see Watson, 1997, pp 446–448.

Incorrect: **SIM plot** – when used to describe a mass chromatogram or an SIM chromatogram. The generation of a mass chromatogram is a different process than the use of the selected-ion-monitoring technique.

Incorrect: **stick plot** – when used to describe a bar-graph mass spectrum.

Incorrect: **total ion chromatogram** – when used to describe an RTICC. It is possible to confuse the total ion chromatogram used as a synonym for RTICC with the result obtained in ion chromatography, or it suggests the direct recording of the total ion current (TIC).

Correct: **total ion current (TIC)** – sometimes used to refer to the sum of all the intensities in a mass spectrum. TIC is synonymous with total ion abundance.

Correct: **mass discrimination** – the ability of an *m/z* analyzer to transmit or detect ions of one *m/z* value more efficiently than another. An instrument can exhibit either *high-* or *low-mass discrimination*. Mass discrimination can be due to instrument design or the contamination condition of the instrument. Transmission quadrupole mass spectrometers exhibit high-mass discrimination that is compensated with either a **Brubaker prefilter**, which is a set of RF-only poles attached to the front of the filter (Brubaker 1968), or a **Turner–Kruger ion-optics lens**, which is a lens that terminates inside the quadrupole field (Barnett 1971).

Correct: **spectral skewing** – used to describe the phenomenon of changes in relative intensities of mass spectral peaks due to the changes in concentration of the analyte in the ion source as the chromatographic component elutes. This phenomenon is not observed in ion trap (quadrupole or magnetic) or time-of-flight (TOF) mass spectrometers. The TOF-MS records all ions sent down the flight tube in a single pulse. The ion trap mass spectrometer records all the ions that have been stored in the trap.

Care must be taken not to confuse spectral skewing with mass discrimination. An example of spectral skewing is shown in **Figure 2**.

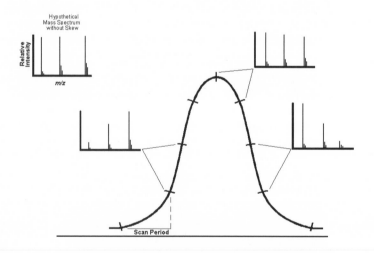

Figure 2. Illustration of spectral skewing.

DATA ACQUISITION TECHNIQUES AND IONIZATION

Correct: **accelerating voltage** – a term used to describe the voltage to transport ions from an ion source to the *m/z* analyzer in a mass spectrometer. This voltage is what gives ions a specific velocity.

Correct: **associative ionization** – a process of ion formation when two excited atoms or molecules in the gas phase interact so that the sum of their internal energies is high enough to produce a single ion that is composed of both reactants. This process differs from **chemi-ionization** in that the resulting ion is composed of the two reactants.

Correct: **atmospheric pressure ionization** (**API**) – causes ions to be in the gas phase at atmospheric pressure as opposed to the reduced pressure that is normally found in the ionization region of an electron ionization (EI) or chemical ionization (CI) mass spectrometer. The term encompasses electrospray and atmospheric pressure chemical ionization.

Incorrect: **API** – when used as an abbreviation for atmospheric pressure chemical ionization (APCI). APCI was originally called atmospheric pressure ionization and abbreviated as API. This use is archaic and causes confusion.

Correct: **atmospheric pressure chemical ionization** (**APCI**) – an ionization technique in which analytes are ionized in ion/molecule reactions that take place at atmospheric pressure as opposed to the 10–100 Pa pressure used when analytes are introduced into the ion source in the gas phase. In APCI, the analytes are pneumatically sprayed into the heated (~400 °C or greater) ion source. The analyte molecules are volatilized. The nitrogen and oxygen molecules in the ion source are ionized with a corona discharge. The oxygen and nitrogen ions react with solvent molecules, which are now in the gas phase, to form the reagent ions. These reagent ions react with the analyte molecules to produce the analyte ions. APCI allows for liquid inlet flow rates of 400–2000 µL min^{-1}. The gas used in this pneumatic nebulization process of the LC eluate is called the **nebulizer gas** In most cases, only single-charge ions are formed. APCI is a mass-dependent technique (i.e., as long as the amount of analyte does not change, the signal will remain the same regardless of the concentration of the solution). "APcI" is an inappropriate abbreviation for atmospheric pressure chemical ionization. This convention has been adopted by Micromass/Waters and should not be proliferated. This technique was developed by a group under Evan Horning at Baylor University, Houston, Texas, in the mid-1970s (Horning 1973). The technique was first commercialized by SCIEX; however, it did not gain popularity as an LC/MS technique until the development of electrospray.

Correct: **atmospheric pressure photoionization** (**APPI**) – an atmospheric pressure ionization technique where ions are formed in the gas phase by photons. In some cases, APPI is carried out with the aid of a dopant that is ionized, which then functions as a reagent ion to bring about ionization of an analyte in an ion/molecule reaction. APPI is another one of the atmospheric pressure techniques used in LC/MS.

Correct: **electrospray (ES)** – a process by which ions that are in solution are caused to be in the gas phase (for subsequent *m/z* analysis) through either a mechanism of ion desorption or ion evaporation. The process is carried out by the production of a fine spray of the solution that contains the analyte from a narrow-bore needle (0.1–0.3 mm diam) to which a potential of 3–5 kV has been applied (6–8 kV in sector instruments). A counter electrode is used to create a field to facilitate the spray formation. The charged droplets evaporate to a point where the number of electrostatic charges on the surface become so large relative to the droplet size that an explosion occurs to produce a number of smaller droplets that also have a surface that contains electrostatic charges. This process repeats until the analyte ion escapes the droplet (ion desorption) or all the solvent has evaporated to leave the ion in the gas phase (ion evaporation). Although the abbreviation **ESI** has been used for *electrospray interface* and *electrospray ionization*, it is believed that the process is not really involved with the formation of ions—just ion desolvation. However, if ionization is any method giving gas-phase ions, then the term could be electrospray ionization. The first time the abbreviation "ESI" is used, it should be defined. The typical flow rate for electrospray is about 1–10 µL min^{-1}. When flow rates of <1 µL min^{-1} are involved (such as in the case of capillary electrophoresis), a make-up solution of the same composition as the mobile phase is introduced through the annular space formed by placing the ES needle in another needle to produce a sheath flow of a nebulizing gas. The ES interface, like the APCI interface, will also have a **curtain gas**, usually a flow of high-purity nitrogen across the orifice to the analyzer region. The curtain gas is used to aid in ion desolvation in ES and is used in ES and APCI to aid in ion declustering and prevention of neutral particles passing from the gas-phase ion-formation region into the *m/z* analyzer. In ES, ions with multiple charges can be formed. Because a mass spectrometer measures the abundance of ions based on their mass to number of charges ratio, ions of very high mass can be detected with conventional *m/z* analyzers (i.e., an ion with 10 positive charges and a mass of 5000 Da would be observed at an *m/z* value of 500 in a mass spectrometer). ES is a concentration-dependent technique (i.e., as the concentration of the analyte in solution decreases, even if the same amount is present, the signal decreases). Although there were several events that contributed to electrospray, its development is credited to John Fenn at Yale University (Whitehouse 1984). The technique gained popularity after Fenn reported the detection of multiple-charge ions in 1988 (Meng 1988). Fenn shared the 2002 Nobel Prize in chemistry for the development of this technique and its application to biological mass spectrometry.

*Incorrect: **spray ionization** – when used as a general term to describe ionization that results from the spraying of a solution into a mass spectrometer; a generic term for APCI, ES, and thermospray. The "MS Terms and Definitions" that appears on the ASMS Web site (http://www.asms.org) under Items of Interest defines spray ionization as "method used to ionize liquid samples directly by electrical, thermal, or pneumatic energy through the formation of a spray of fine droplets". This term is too encompassing and can result in some degree of confusion.

Incorrect: **IonSpray**™ **(ISP)** – when used as a synonym for pneumatically assisted electrospray. IonSpray is the trade name used by PE/SCIEX to describe pneumatically assisted electrospray. The pneumatically assisted electrospray technique allows for liquid flow rates of 10–1000 µL min^{-1}. IonSpray was coined and developed by Jack Henion at Cornell University, Ithaca, New York (Bruins 1987). Today, all commercially available ES mass spectrometers have a pneumatic assist. IonSpray should only be used to describe the trademarked product for PE/SCIEX.

Correct: **microelectrospray (micro-ES)** – the low-flow electrospray technique that does not use a make-up solvent where a pump is used to establish the sample flow. Microelectrospray is the first low-flow-rate ES technique described, but it was referred to as "nanoliter flow LC/ES/MS" in its first presentation (Emmett 1993). The terms "micro-ES" (Emmett 1994) and "microES" used by Matthias Wilm and Matthias Mann to describe what later became known as "nano-ES" (Wilm 1994) were used for the first time at the 1994 ASMS Conference in Chicago. The Wilm et al. system used a gas pressure to cause the flow, whereas the Emmett et al. system used a mechanical pump. In a subsequent publication, the reported flow rates for micro-ES were 300–800 nL min^{-1} (Emmett 1994[2]).

Correct: **nanoelectrospray (nano-ES)** – the low-flow electrospray technique that does not use a make-up solvent where the sample flow is dependent on the potential on the tip of the electrospray needle and/or a gas pressure to push the sample through the needle. The nanoelectrospray name for this technique was originally coined by the developers, Matthias Mann and Matthias Wilm (Wilm 1996), at the European Molecular Biology Laboratory (EMBL) in Heidelberg, Germany. Mark Emmett differentiated the techniques of nano-ES and micro-ES from one another by virtue of whether or not a mechanical pump is used (Emmett 1997). The nano-ES technique also differs from micro-ES in the diameter of the needle (~1–3 µm compared to ~50 µm for micro-ES) and the flow rates (~25 nL min^{-1} for nano-ES as compared to 300–800 nL min^{-1} for micro-ES). **NanoSpray**™ is the Bruker Daltonics trade name for nanoelectrospray. This electrospray technique has also been referred to as **MicroIonSpray**™ and **NanoFlow**™ **ES** (trade names that refer to the technique as marketed by PE/Sciex and Micromass, respectively). These three trade names should only be used to describe the trademarked products.

Note: Before using a term to describe an electrospray or a low-flow electrospray technique, care must be taken to make sure that the actual process is clearly understood, and the process is not supposed to be implicit by the name.

Correct: **charge-reduction electrospray mass spectrometry (CREMS)** – a technique used to reduce the number of charge sites on a multiple-charge ion produced by electrospray. Although this process has been reported by at least one other investigator (Stephenson 1997), the term was coined in relation to the use of a polonium-210 particle source (Scalf 1999). The technique makes the measurement of mass as a function of charge site easier while taking advantage of the electrospray process.

Correct: **coldspray (CS)** – a variant of the electrospray interface that operates so that the temperature of the drying gas and capillary are maintained below –20 °C in order to promote ionization based on increased polarizability of compounds resulting from a high dielectric constant at these low temperatures. This technique was originally developed for the ionization of thermally unstable ionic metal complexes. These substances exhibited extreme decomposition when analysis was attempted using conventional electrospray. This technique was developed by Kento Yamaguchi (Sakamoto 2000) at the Chemical Analysis Center in Chiba University in Chiba, Japan.

Correct: **desorption electrospray ionization (DESI)** – a technique for the analysis of analytes on surfaces of types related and unrelated to the analyte. DESI involves directing the charged liquid droplets and the gas jet from a pneumatically assisted electrospray ion source at the surface to be analyzed. The analytes are desorbed into the gas phase by electrostatic and pneumatic means with gas-phase ions being directed to an atmospheric sampling orifice of a mass spectrometer. The exact mechanism of desorption and ionization has not been explained. The technique was developed by R. Graham Cooks (Takáts 2004) at Purdue University, West Lafayette, Indiana.

Correct: **direct liquid inlet (DLI) probe** – a device to introduce a liquid into the ion source of a mass spectrometer. This probe is most often used as a liquid chromatography interface for an electron ionization (EI) or chemical ionization (CI) source.

Correct: **particle beam (PB) interface** – a technique used in LC/MS by which analyte molecules are desolvated in a heated chamber and passed through a momentum separator to remove the solvent vapor from clumps of moist molecules. These clumps of moist molecules enter a conventional EI or CI source, where they are volatilized by impacting onto a heated splatter plate. The PB interface used with EI is the only LC/MS technique that produces a conventional EI mass spectrum. Waters Corporation uses the trademarked term **ThermaBeam**™ to describe this interface technique. **MAGIC** (monodisperse aerosol generation interface for combined LC/MS) was the name applied to this technique (Willoughby 1984) and was used commercially by Hewlett-Packard (now known as Agilent Technologies).

Correct: **thermospray (TSP)** – an LC/MS technique operated in a "filament-on" or "filament-off" mode. In the *filament-on* mode, analyte ions are formed in a reduced pressure environment through the direct ionization of the mobile phase with an electron beam or corona discharge followed by ion/molecule reactions between the nascent reagent ions and the analyte molecules in the gas phase. In the *filament-off* mode of operation, thermospray produces ions in the gas phase through a process similar to ES (i.e., ions in solution are caused to enter the gas phase for subsequent m/z analyses). The ion evaporation or desolvation in thermospray is accomplished at a reduced pressure and without the aid of a high potential. This technique is no longer commercially available. Thermospray was developed by Marvin Vestal at the University of Houston, Houston, Texas (Blakely 1983).

Correct: **chemi-ionization** – a process of ionization that comes about when gas-phase molecules come in contact with internally excited gas-phase molecules of the same or different species. This process results in an ion of the molecule, not an additive product as in **associative ionization**. This term is *not* synonymous with chemical ionization.

Correct: **chemical ionization (CI)** or **positive-ion CI (PICI)** – ionization of an analyte that occurs as a result of an ion/molecule reaction. A **reagent gas** (sometimes referred to as the **conjugate base** of a positive reagent ion) is ionized by an electron beam. The resulting ions react with neutral molecules of the reagent gas to form reagent ions (e.g., CH_5^+ from $CH_4^{+\bullet} + CH_4$). These reagent ions react with analyte molecules to produce analyte ions. The typical CI source pressure in beam-type instruments is 10–100 Pa. The reagent gas partial pressure in an internal ionization ion trap mass spectrometer is $\sim 10^{-3}$ Pa. Chemical ionization was developed by Burnaby Munson and Frank Field at the University of Texas, Austin, Texas (Munson 1966). There are four different processes that produce positive ions from chemical ionization:

charge transfer (The reagent ion, missing an e^-, takes an e^- from the analyte.)
$$CH_4^{+\bullet} \; + \; RH \longrightarrow RH^{+\bullet} \; + \; CH_4 \qquad M^{+\bullet}$$

proton transfer (Common when the analyte molecule has a higher proton affinity (PA) than the reagent gas. The analyte takes H^+ from the reagent ion.)
$$CH_5^+ \; + \; RH \longrightarrow RH_2^+ \; + \; CH_4 \qquad [M + 1]^+$$
$$C_2H_5^+ \; + \; RH \longrightarrow RH_2^+ \; + \; C_2H_4$$

hydride abstraction (The reagent ion has a high hydride affinity, which is the ability to remove a H^- from the analyte molecule.)
$$CF_3^+ \; + \; RH \longrightarrow R^+ \; + \; CF_3H \qquad [M - 1]^+$$
$$C_2H_5^+ \; + \; RH \longrightarrow R^+ \; + \; C_2H_6$$

collision-stabilized complexes (Occurs when the PA of the analyte and reagent gas are comparable. The reagent ion becomes attached to the analyte. When methane is used, the $[M + 1]^+$, $[M + 29]^+$, and $[M + 41]^+$ series is a very good confirmation of the nominal mass of the molecule.)
$$C_2H_5^+ \; + \; RH \longrightarrow [C_2H_5:RH]^+ \qquad [M + 29]^+$$
$$C_3H_5^+ \; + \; RH \longrightarrow [C_3H_5:RH]^+ \qquad [M + 41]^+$$

Correct: **negative-ion CI (NICI)** – refers to the formation of negative ions by the reaction between an analyte molecule and an anion that was formed from a reagent gas.

negative ion/molecule reaction
$$AB \; + \; C^- \longrightarrow ABC^- \text{ or } [AB - H]^- + HC$$

Incorrect: **negative-ion CI (NICI)** – when used to describe the formation of negative ions by the analyte molecule's capture of a low-energy electron. The reason this process has been called negative-ion CI is that the low-energy electrons can be produced by the ionization of a gas (usually methane) under high-pressure conditions similar to those used in conventional CI when methane is used as the reagent gas.

Correct: electron capture/negative-ion (ECN) detection, electron capture/negative ionization (EC/NI), or **resonance electron capture ionization (RECI)** – refers to producing negative ions by the reaction of **thermal electrons** with molecules in a mass spectrometer. Thermal electrons are low-energy electrons that have average kinetic energies that are the same order-of-magnitude as the thermal energy of the molecules that they are intended to ionize. These terms are the correct terms to describe the formation of negative ions when an analyte molecule captures a low-energy electron. This process, first reported as a GC/MS technique by Ralph Dougherty (Dougherty 1972) and later by Don Hunt and George Stafford (Hunt 1976), usually happens under high-pressure conditions that are similar to those used in conventional CI. The resulting negative molecular ions are symbolized as $M^{-\cdot}$, which indicates an odd-electron negative ion. The results of electron capture ionization are as follows.

resonance electron capture
$$AB \;+\; e^- \;(\sim 0.1\,eV) \longrightarrow AB^{-\cdot}$$
dissociative electron capture
$$AB \;+\; e^- \;(0\text{–}15\,eV) \longrightarrow A^{\cdot} \;+\; B^-$$
ion-pair formation that results from electron capture
$$AB \;+\; e^- \;(>10\,eV) \longrightarrow A^- \;+\; B^+ \;+\; e^-$$

Correct: **pulsed-positive negative-ion chemical ionization (PPNICI)** – a data acquisition method used primarily with transmission quadrupole mass spectrometers by which alternating spectra are taken of positive ions formed by CI and negative ions formed by CI or electron capture ionization. The hardware configuration for PPNICI was under patent issued to George Stafford, a graduate student at the University of Virginia under Donald F. Hunt, and licensed to the Finnigan Corporation (Hunt 1976).

Correct: **tunable energy electron monochromator (TEEM™)** – a device using an electric and a magnetic field to allow the energy of electrons produced by a filament to be tuned to desired energy. These electrons are then used in EC/NI. Use of the TEEM allows the energy of the electrons to be tuned to that of target analytes, resulting in more stable and reproducible signals (Larameé 2000).

Correct: **chemical reaction interface mass spectrometry (CRIMS)** – a process in which the eluate from a chromatographic process (LC or GC), which has been enriched with respect to the analyte, is passed into a microwave reaction chamber that contains a reaction gas selected to produce a simple product gas such as CO_2, NO, SCI, (etc.) from organic analytes and their metabolites that contain stable isotopically labeled analytes. A measure of the ratio of naturally occurring product gas to isotopically labeled product gas indicates the presence of metabolites (Abramson 1994).

Correct: **direct analysis in real time (DART**™**)** – an ionization technique that operates at atmospheric pressure and does not involve any electrical fields to form ions of analytes on the surface of all types of materials (gases and liquids) as well as gases. The technique uses excited-state species such as helium metastable atoms (2^3S) that have an energy of 19.8 eV. These excited-state species react with atmospheric water to produce protonated clusters $[(H_2O_{n-1})H]^+$ and hydroxyl ions. The presence of dopants will produce adduct ions and, in some cases, increase sensitivity. The protonated water clusters can produce protonated molecules of analytes. The system can also produce thermalized electrons to form negative-charge molecular ions. Gases such as N_2 will produce positive-charge molecular ions through Penning ionization. Other ionization systems using corona discharge in air, as is used by DART, produce nitrogen oxide that interfere with the analysis of nitrogen-based explosives. DART has been shown to be effective at the detection of ClO_3^- and ClO_4^- ions from sodium perchlorate deposited from solution on a glass rod as well as detection of volatile solvents by opening a bottle with a material such as acetonitrile 10–20 feet from the open ionization source. Various dopants have been used to enhance analyses by the detection of adduct ions like $[M + NH_4]^+$ when vapors from NH_4OH was used. Unlike ES, alkali adducts are never seen in conjunction with protonated molecules. Ionization takes place without the need for any solvents or sample preparation. This technique has been used with a high-resolution TOF mass spectrometer. It was introduced by JEOL at the 2005 ASMS Sanibel Conference (Cody 2005).

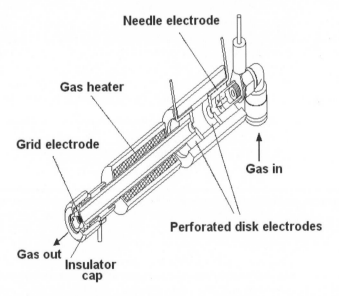

Figure 3. Schematic representation of the DART ionization device.

Correct: **desorption/chemical ionization** (**DCI**) – a technique by which a nonvolatile thermally labile analyte produces a CI mass spectrum. The analyte is dissolved in a volatile solvent. A drop of the resulting solution is put on a highly conductive loosely coiled element, and the solvent is evaporated. The element, now containing only the analyte, is introduced into a conventional ion source of a mass spectrometer via an insertion probe. In the presence of CI reagent ions at 10–100 Pa, a current is passed through the element. Analyte ions produced from ion/molecule reactions are desorbed from the surface of the element into the gas phase for *m/z* analysis. This technique has produced mass spectra of sugars without the need for derivatization.

Correct: **desorption/ionization** (**DI**) – describes the ionization of an analyte in a solid matrix or solution followed by the subsequent desorption of the ions into the gas phase for *m/z* separation and detection.

Correct: **direct infusion** – a sample introduction technique. It is the introduction of a liquid sample into the area of the mass spectrometer where ions are brought into (or produced in) the gas phase. Direct infusion results in a continuous concentration of analyte in the ion source/interface of the mass spectrometer.

Correct: **direct insertion probe** – one of two types of probes used to introduce solids into the EI or conventional CI ion source of a mass spectrometer. A sample is placed in a small glass tube (similar to a melting-point capillary tube) that is about 10 mm in length. The tube is then fitted to the end of a probe and is inserted into an interface mounted on the ion-source housing of the mass spectrometer. The tip is evacuated, a vacuum interlock is opened, and the probe is pushed into the ion source to position the tube containing the sample near the electron beam. The tip of the probe is then heated to volatilize the sample. This type of probe can easily contaminate the mass spectrometer because of the volume of sample delivered to the ion source. This probe is sometimes referred to as a **solids probe**.

Correct: **direct exposure probe** – the other type of probe used to introduce solids into the EI or conventional CI ion source. A sample is dissolved in a solvent. A drop of the solution is placed on the end of the probe (usually a rounded glass tip). The solvent is evaporated leaving a thin film of uniform thickness on the inside of the probe. The probe is inserted into the ion-source housing using the same mechanisms as the direct insertion probe. The tip is then heated to volatilize the analyte. This technique is preferred for the introduction of solids because it will produce far less contamination. A variation of this probe is used with desorption/chemical ionization (DCI). The rounded glass tip is replaced with a filament. The sample solution is then put on the filament, the probe is inserted into the ion-source housing, and a current is passed through the filament to aid in the production of gas-phase ions (see **desorption/chemical ionization**).

Correct: **data-dependent acquisition** or **experiment** (**DDA** or **DDE**) – a process of data acquisition that causes the acquisition parameters to change from spectrum to spectrum based on information automatically derived from the previous spectrum. DDA is often employed for obtaining MS/MS data, used in a given chromatographic elution, based on the base peak in the previous spectrum. DDA is a Thermo Electron term. Waters uses **data-directed acquisition**. Applied Biosystems/MSD Sciex uses the term **information-dependent data acquisition** (**IDA**). Other companies may use other terms.

Correct: **electron ionization** (**EI**) – ionization of analyte molecules in the gas phase (10^{-1} to 10^{-4} Pa) by electrons accelerated between 50 and 100 V. The original standard was 70 V. After the development of chemical ionization as a mass spectrometry technique (1966), papers appeared that used electron ionization as opposed to chemical ionization. Previously, EI was used as an abbreviation for **electron impact**. The use of electron impact is the reason why "EI ionization" is seen in print. The use of "electron impact" or "EI as an abbreviation for electron impact" is **incorrect**. The *Current IUPAC Recommendations* states, "Electrons and photons do not 'impact' molecules or atoms. They interact with them in ways that result in various electronic excitations, including ionization." Also, Ken Busch points out, "An electron is far too light to transfer kinetic energy to a sample molecule in a collision process" (Busch 1995). Although many refinements were made that led to the current EI source designs, the original development is attributed to Arthur Jeffrey Dempster (1886–1950), who developed the electron bombardment source at the University of Chicago, Chicago, Illinois (Dempster 1922).

Correct: **field desorption** (**FD**) – an ionization technique by which analyte ions are desorbed from the surface of one of the two electrodes used to produce an electrical field of 10^7 to 10^8 V cm^{-1}. FD is the first desorption technique to be seriously considered in mass spectrometry. FD uses specially prepared sample emitters that allow for the production of protonated molecules (MH$^+$). FD has been used for the analysis of thermally labile nonvolatile samples. This technique has been largely replaced by ES. Formation of ions via an electrical field had its origin with field ion microscopy developed in 1951 by E. W. Müller in Germany. The first developments involving FD were by R. Gomer and M. G. Inghram (Gomer 1954, 1955; Beckey 1977).

Correct: **field ionization** (**FI**) – an ionization technique that results in a high abundance of molecular ions. Ionization of an analyte molecule in the vapor phase takes place in an electrical field (10^7 to 10^8 V cm^{-1}) maintained between two sharp points or edges of two electrodes. The technique of FI was developed by H. D. Beckey, Institut für Physikalische Chemie der Universität Bonn, Bonn, Germany, in 1957 (Beckey 1963, 1977).

Correct: high-**field asymmetric** waveform **ion mobility spectrometry** (**FAIMS**) – FAIMS takes advantage of differences in an ion's mobility under high and low electric field strengths. An ion mixture is introduced between two electrodes in FAIMS. A gas flow carries the ions from the ion inlet to the *m/z* analyzer. Selected subsets of ions can be transmitted to the *m/z* analyzer by adjusting the voltage applied to the electrodes. Interfering ions that are not transmitted at a specific voltage collide with electrodes. The concept of high-field ion mobility was first reported in Russia (Buryakov 1991, 1993) and was shortly later extensively researched at the National Research Council of Canada, which led to its commercialization (Purves 1998). FAIMS has shown a remarkable ability to produce significantly higher-quality spectra generated by electrospray and APCI.

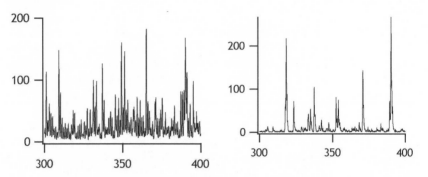

Figure 4. The right panel is a spectrum of a mixture of a tryptic digest of several proteins (concentration 500 fmol μL^{-1}) without the addtion of FAIMS. The left panel is a spectrum of the same sample using FAIMS.
(Courtesy of Ionalytics Corporation, Ottawa, Canada)

Correct: **flow injection** – a process by which a sample is injected into a continuous liquid flow that enters into the mass spectrometer. Sample introduction is the same as a liquid chromatograph injection; however, in flow injection, there is no column between the injector and the mass spectrometer. The flow injection technique is sometimes called **loop injection** because the sample is loaded into a loop on a sampling valve rather than being injected into a flowing stream with a syringe. Flow injection is sometimes referred to as *flow injection analysis* (FIA).

Incorrect: **flow injection** – when used to refer to the technique of direct infusion.

Correct: **fast atom bombardment (FAB)** – **FAB mass spectrometry** – a desorption/ionization technique that is used to obtain ions of large nonvolatile thermally labile analytes in the gas phase for *m/z* analyses. A solution of the analyte is mixed with a matrix (usually glycerol) and placed in an evacuated ion-source housing. The surface of this mixture is bombarded with a stream of atoms (usually argon) that have been given approximately 5–8 kV of translational energy. The impact of the particles on the surface produces molecular sputtering, which results in desorption of ions from the liquid/vacuum interface into the gas phase. The analyte may be in ionic form in the target sample, or it is ionized by proton transfer that accompanies the bombardment. A related technique, **liquid secondary-ion mass spectrometry (LSIMS)**, employs a stream of ions (often cesium) to accomplish the ionization. FAB and LSIMS have largely been replaced by ES and MALDI. The FAB technique is attributed to Mickey Barber, University of Manchester Institute of Science, Manchester, England (Barber 1981).

Correct: **continuous-flow FAB** (**CF-FAB**) – largely an LC/MS technique (but could be used to improve signal-to-background in infusion experiments) in which the eluate from the liquid chromatograph is continually mixed with the FAB matrix material (usually glycerol), and the mixture passes through a needle with a slanted tip. The bombardment is pulsed to produce each mass spectrum as a continually changing sample flows to the FAB target. This technique has been largely replaced by ES. The development of CF-FAB is jointly attributed to Richard Caprioli at the University of Texas Medical School, Houston, Texas (Caprioli 1986); and M. Ito, who developed Frit-FAB in Japan (Ito 1985).

Correct: **ionization cross section** – a measure of the probability that ionization of an atom or molecule will occur in an electron ionization or photoionization process. The higher the ionization cross section, the higher the probability that ionization will occur.

Correct: **ion mobility spectrometry** (**IMS**) – the separation of ions of different m/z values based on their mobility differences. A gaseous ion's mobility (**ion mobility**) is the ratio of an ion's velocity at atmospheric pressure to an applied electric field. The ion accelerates through the field until it collides with a **drift gas** (usually N_2 or air), which is flowing in a direction opposite (to minimize the number of collisions between the molecules and the ion) to the increasing field strength produced by a series of **guard rings** found in the **drift tube**. IMS is capable of measuring very low ion currents (below 10^{-12} A), and it has a moderate ability to separate ions of differing m/z values. It has been used as a standalone device and a gas chromatographic detector. When the technique was first introduced in the early 1970s, it was called **plasma chromatography** and **gaseous electrophoresis**. Because the terms *electrophoresis* and *chromatography* had to be expanded beyond their means, the term *ion mobility* (because that was the property being exploited) *spectrometry* (because mass spectrometry was a well-accepted use of the word *spectrometry*) was adopted (Hill 1990).

Correct: **inductively coupled plasma** (**ICP**) – an ionization technique by which analyte ions are formed from gas-phase atoms that are produced by the nebulization of a solution containing ions of the analyte into a plasma, which results from an electroless discharge in the argon gas that is driven by energy coupled from an RF generator. The temperature of the plasma is about 8000 °C. This ionization technique is used for the mass spectral identification of inorganic substances, especially heavy metals.

Correct: **glow discharge ionization** (**GDI**) – an ionization technique by which analyte ions are produced from mainly inorganic samples by the collision of accelerated ions with a cathode that is made of or coated with the solid analyte to produce gas-phase atoms that are subsequently ionized by electrons produced by an electrical discharge in an inert gas. Nonconducting analytes are mixed with a conducting material and coated onto the cathode. The accelerated ions are produced at a pressure of $10–10^3$ Pa of gas that is between the anode and the cathode. The ions accelerated to the surface are those produced by high-energy electrons in the "cathode dark space" (the area between the cathode surface and the glow produced by low-energy electron ionization of the gas). The atomization of the sample (sputtering) results in gas-phase analyte molecules and a few gas-phase ions. The gas-phase molecules are then ionized by high-energy electrons in the glow region (Harrison 1986; Barshick 2000). Nearly all elements can be analyzed by glow discharge.

Correct: **gas chromatography/mass spectrometry (GC/MS)** is the technique by which a mixture of analytes is separated into its individual components by gas chromatography. The chromatographic eluate passes into the ionization source (usually EI, CI, FI, or ECNI) of the mass spectrometer. If the flow rate (a function of the linear velocity of the carrier gas) is too great for the vacuum system of the mass spectrometer, then one of several devices can be used to enrich the eluate of the column or divert a portion of it from the ion source. These devices are the **jet separator**, which was refined by Ryhage (Ryhage 1964) from the device first described by Becker (Becker 1957); the **membrane separator** developed by Llewellyn and Littlejohn, a.k.a. the **Llewellyn separator** (Llewellyn 1966); the **molecular effusion separator** developed by Watson and Biemann, a.k.a. the **Watson–Biemann separator** (Watson 1964); and the **open-split interface** introduced by Finnigan Corporation on the ITD 700 quadrupole ion trap GC-MS in 1984. Only the jet separator was widely used with commercial gas chromatographs. Hewlett-Packard did provide the Llewellyn separator on its commercially distributed instruments for a brief period before changing to the jet separator. All of these devices were designed to compensate for the flow rates of 20–30 mL min^{-1} produced from packed GC columns (glass tubular columns, ~2 mm i.d., with a specific geometry for the instrument in which they were used, and were packed with particles of an inert substance [usually diatomaceous earth] coated with a high-boiling viscous liquid). Once these wall-coated open-tubular (WCOT) capillary columns constructed of fused silica became generally accepted, there was no longer a need for such devices. The instrument's vacuum system can easily handle the carrier gas that is diluting the gas-phase analyte. With these WCOT columns, the concentration of the analyte is much higher in the eluate than it was with the packed columns. A good explanation of these devices and how they function can be found in *Introduction to Mass Spectrometry*, 3rd ed. (Watson 1997).

Correct: **laser microprobe mass analyzer (LAMMA)** – the first name given to a commercial instrument using laser desorption (LAMMA-500 Leybold Heraeus, Koln, West Germany). This instrument was built on the work published by Franz Hillenkamp (Hillenkamp 1975; Wechsung 1978; Denoyer 1982). This was the pioneering work that led to MALDI.

Correct: **liquid chromatography/mass spectrometry (LC/MS)** – the technique by which a mixture of analytes is separated into its individual components by liquid chromatography. The chromatographic eluate (a solution containing the analyte and mobile phase) passes into the mass spectrometer's interface. Like the situation that existed with packed GC columns, it is necessary to remove the mobile phase (a liquid in LC/MS) from the analyte; and it is also necessary to get analyte ions into the gas phase. Both of these steps are accomplished by various LC/MS interface and ionization techniques.

Correct: **matrix-assisted laser desorption/ionization** (**MALDI**) – a desorption/ionization technique in which a laser is used to produce ions from analytes that are present in a solid matrix. Matrix molecules undergo electronic excitation with a UV laser or vibrational excitation with an IR laser—clearly a process yet to be understood. The matrix acts as an energy disperser to bring about the ionization of the analyte. The laser provides energy for the ionization and the desorption of the ions from the matrix. Unlike other mass spectrometry techniques where samples are introduced in the region where gas-phase ions are obtained, MALDI samples are introduced on sample holders with indentations called *wells*. The sample consists of the MALDI matrix and the analyte mixed together as in a contiguous solid state. The well containing the sample is referred to as the **MALDI target**. The MALDI target is sometimes arranged as a **microtiter plate** that has 96, 384, (etc.) wells, which allows for automated preparation of the MALDI targets. This technique produces predominantly single-charge ions; however, some double- and triple-charge ions may be observed. MALDI was developed by Franz Hillenkamp and Michael Karas at the Institute für medizinische Physik und Biophysik, Universität Münster, Münster, Germany (Karas 1987; Hillenkamp 1991).

According to Ron Beavis (Beavis 1992), the term "matrix-assisted laser desorption" was coined by Hillenkamp and Karas in the title of a presentation at the 10th International Mass Spectrometry Conference held in Swansea, United Kingdom, 9–13 September 1985, "Matrix Assisted UV Laser Desorption of Non-volatile Organic Compounds" (Karas 1986).

MALDI is different from Tanaka's technique, which he called "ultra-fine metal plus liquid matrix method" or "soft laser desorption" (SLD), the designation used by the 2002 Nobel Prize Committee (see **soft laser desorption**: **SLD**). After the initial presentation of SLD, the technique was never pursued. MALDI has become one of the ubiquitous techniques used in biological mass spectrometry. Although the Hillenkamp and Karas MALDI technique dates to 1985 or before, they did not show the analysis of a >10,000-Da protein until almost a year after Tanaka's presentation (1988 International Mass Spectrometry Conference held in Bordeaux, France, 29 August–2 September).

A variation of the MALDI technique is **atmospheric pressure MALDI** (**AP MALDI**). Vacuum MALDI is usually performed with a time-of-flight mass spectrometer. The mainly single-charge ions produced will have an initial spatial spread and an initial kinetic energy, which results in broad peaks with poor resolution. In AP MALDI, ions are produced at atmospheric pressure; and an ion beam is culminated with a combination of orifices, ion guides, and skimmers into a storage area where they are extracted via an orthogonal pulse into a TOF mass spectrometer or analyzed using a quadrupole ion trap. The use of the QIT allows for MS^n analyses of the ions produced by MALDI. The technique had its commercial introduction in 2001 from Mass Technologies as an after-market product for the Finnigan LCQ QIT mass spectrometer with an ES and/or APCI interface for LC/MS.

affinity mass spectrometry using **surface-enhanced laser desorption/ionization** (**SELDI**) is a variation of the MALDI technique. Affinity mass spectrometry and SELDI were pioneered by T. William Hutchens at Baylor College of Medicine in Houston, Texas (Hutchens 1992, 1993). The technique uses a MALDI plate that is first treated with an antibody that will cause the immunoaffinity to capture the antigens in complex biological matrices. The other biological substances that would ordinarily interfere with the mass spectrometry of the antigen are removed by washing, leaving the purified antigen bonded to the surface. The technique of affinity mass spectrometry is described by Randall Nelson and co-workers in **Figure 5** (Nelson 1997).

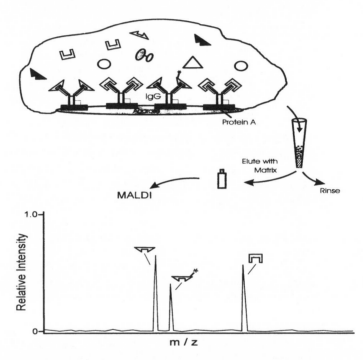

Figure 5. Illustration by Nelson and co-workers, showing how antigens and chemically modified variants used in quantitation are captured onto the antibodies in the affinity probe. (Nelson, RW; Krone, JR; Bieber, AL; Williams, P "Mass Spectrometric Immunoassay" *Anal. Chem.* **1995**, *67*, 1153–1158)

Correct: **desorption/ionization from silicon** (**DIOS**) – a variant of the SELDI technique where porous silicon plates are placed on a MALDI plate. The plates contain photopatterned spots or grids prepared through illumination of *n*-type silicon with a 300-W tungsten filament through a mask and an *f*/50 reducing lens. The sample is placed on the porous silicon plate and allowed to dry, followed by laser-induced desorption/ionization mass spectrometry. This technique was developed by Gary Siuzdak at the Scripps Research Institute in La Jolla, California (Wei 1999).

Correct: **membrane introduction mass spectrometry** (**MIMS**) – a process where nonionic nonpolar analytes dissolved in water are adsorbed from solution by a membrane that is not permeable to the water. The analytes pass through the membrane into the mass spectrometer. By the continuous monitoring of ions specific for various analytes, a quantitative determination can be made. This technique has been used in a continuous monitoring process of the volatile compounds in water. Any of the semivolatile compounds that pass through the membrane can also be monitored. It can also be used as a substitute for a solid probe, provided the analyte will pass through the membrane (Bauer 1993).

Correct: **modes of data acquisition** – mass spectrometers are operated according to differences in the *m/z* values of ions that can be separated. Instruments are operated at **high resolution** (meaning a resolving power [R] M/ΔM ≥10,000), **medium resolution** (R ≥5000 and <10,000), or **low resolution** (where the **resolution** [ΔM] is ≅ 1). A common mistake made in referring to instruments in the high-resolution mode is that they automatically produce **measured accurate mass** data. All instruments are capable of yielding measured accurate mass data regardless of their resolving power capability through the use of **peak matching**. An instrument capable of high resolving power is necessary to assure that the *m/z* value of the ion being measured is of an ion of a single elemental and/or isotope composition. The accurately determined *m/z* value of a mass spectral peak may be a weighted average of multiple ions of different isotope and/or elemental compositions if the **resolving power** of the instrument used is insufficient to separate very small differences in *m/z* values.

Correct: **open access** – a mass spectrometric operational technique by which individual users place samples on an autosampler, request a specific analytical method from a computer menu, and receive results via electronic communications. This eliminates the necessity for submission of sample to a mass spectrometry facility and reduces the time necessary for the user to receive data, such as molecular mass information.

Correct: **photoionization** (**PI**) – an ionization technique by which analyte ions are produced from the reaction of analyte molecules with photons. This is a soft ionization technique resulting in spectra similar to 10-eV electron ionization spectra.

Correct: **plasma desorption** (**PD**) – a technique that uses the fission products of ^{252}Cf to produce ions of large-molecular-weight nonvolatile analytes. PD uses a TOF-MS as the *m/z* analyzer. An average spectrum results from several hours of data acquisition. The technique was developed by Robert Macfarlane at Texas A&M University, College Station, Texas, in the mid-1970s (Sundqvist 1985). The technique is no longer in use and was replaced by ES and MALDI.

Correct: **soft laser desorption** (**SLD**) – the term used by Karin Markides in her report on the 2002 Nobel Prize in chemistry (Markides 2002) to describe the technique originally referred to as "ultra-fine metal plus liquid matrix method" by Koichi Tanaka of the Shimadzu Corporation to obtain a mass spectrum showing a protonated molecule peak for a pentamer of lysozyme. The sample of 10 µg per 10 µL in distilled water was placed on a sample holder. A suspension of 300-Å-diameter cobalt powder in glycerin was dissolved [sic] in an ethanol/acetone solution. 10 μL of this slurry was added to the sample holder. The resulting mixture was then vacuum dried to remove the volatile components. Several one-shot spectra were accumulated using a nitrogen laser (wavelength: 337 nm) (Tanaka 1987, 1988). Tanaka shared the 2002 Nobel Prize in chemistry as a result of this single experiment. Soft laser desorption appears to be an artificial term invented by the Royal Swedish Academy of Science, 2002, and had never previously appeared in the literature.

Correct: **soft ionization** – a term applied to ionization techniques that produce ions representing the intact molecule that have low internal energy resulting in little or no fragmentation. Examples are CI, MALDI, FAB, FD, PD, DCI, APCI, and ES.

Correct: **spark (source) ionization** – an ionization technique by which analyte ions are formed when a solid sample is vaporized in an intermittent electric discharge (a spark). IUPAC recommends that this technique not include the word "source" in its description.

Correct: **thermal ionization (TI)** – an ionization technique by which analyte ions are produced by heating molecules on a surface. Both the ions and the neutral particles will be evaporated. Positive and negative ions can be formed by this process. Thermal ionization is primarily used in the elemental analysis. It has extremely high sensitivity, requiring small sample amounts. Under favorable conditions, picogram-size samples can be analyzed. Nearly every element that exists as a solid has been analyzed using thermal ionization. The technique is used in isotope studies; however, especially for lighter species, preferential evaporation of different mass isotopes of an element can take place. This is one of the earliest ionization techniques used in mass spectrometry.

Correct: **vertical ionization** – a process by which a positive ion is formed by the rapid removal of an electron from a ground-state or excited molecule so that no change occurs in the positions or momenta of the atoms. These ions are often in an excited state.

*Correct: **collisionally activated dissociation (CAD)** and **collision-induced dissociation (CID)** – terms that describe ion fragmentation in an MS/MS experiment or in the region between the atmospheric pressure part of a mass spectrometer and the area where the *m/z* analyzer is located. The precursor ion has its translational energy converted to internal energy by inelastic collisions* with neutral molecules to bring about a dissociation. CAD is used interchangeably with CID. An ion can undergo collisional activation or collisional excitation without fragmenting. With two terms used to describe the same event, confusion can result. Because ions can be collisionally activated or be in a state of collisional excitation, *collisionally activated dissociation* is the preference of this author. The two terms are considered equal in the *Current IUPAC Recommendations*. Graham Cooks and John Beynon are credited with coining the term

---

* An inelastic collision occurs when two objects in motion collide with one another so that the translational motion is converted into internal energy. An elastic collision occurs when the internal energy and the sum of the translational energy of the two objects remain unchanged. Only the direction of the two objects changes.

"collision-induced decomposition" (CID) (Cooks 1973) as the process in MS/MS. Beynon refers to "collision-induced reactions" in his classic mass spectrometry text (Beynon 1960). Klaus Biemann cites the 1957 works on collision-induced dissociation of H. M. Rosenstock and C. E. Melton (Rosenstock 1957) in his classic mass spectrometry treatise (Biemann 1962). Fred McLafferty is credited with the term "collisionally activated dissociation" (CAD) (McLafferty 1983) and used the term "collisional activation" in a seminal 1973 presentation (McLafferty 1973). According to McLafferty in the Glossary and Abbreviations section in the front of the 1983 *Tandem Mass Spectrometry* book, "CAD is preferred to CID, as recommended by the 1979 Washington, DC, ACS meeting discussion chaired by R. G. Cooks."

Correct: **electron-transfer dissociation** (**ETD**) – a process by which an electron is transfered to a multiple protonated molecule of a peptide or protein, resulting in an odd-electron ion that fragments in such a way as to preserve posttranslational modification. This is usually carried out in a 3-D or linear QIT (Syka 2004). The results of ETD are the same as reported for **electron capture dissociation** (**ECD**) of peptides and proteins in an FTICR-MS using a dense population of near-thermal electrons (Zubarev 1998).

Correct: **ion/molecule reaction** – describes an ion/neutral reaction where the neutral species is a molecule.

Incorrect: **ion-molecule reaction** – when used to suggest a reaction between an ion and a molecule. Because of the presence of the hyphen, this term suggests that a species that is both an ion and a molecule is being described. Therefore, *ion-molecule reaction* describes a different event than *ion/molecule reaction*.

Correct: **mass analysis** – a process by which a mixture of ionic or neutral species is identified according to the mass-to-charge ratios (ions) or their aggregate atomic masses (neutrals). The analysis may be quantitative and/or qualitative.

Incorrect: **mass spectroscopy** – when used to imply the use of an optical device. This conclusion is reached based on the definition of spectroscopy found in any dictionary. There are no light sources in mass spectrometry. Photoionization has been reported; however, it is not an often-used means of producing ions in a mass spectrometer ($M + h\nu \rightarrow M^{+\cdot} + e$). Mass spectroscopy has been used in a loose sense to include the use of mass spectrometers (abundance positive-ray analysis) and mass spectrographs (accurate-mass positive-ray analysis) as well as the studies of isotopic abundance, precise mass determination, analytical chemical use, appearance potential, and so forth. Mass spectroscopy is too encompassing for general use. Although declining in use, mass spectroscopy is still a popular term, especially with a large segment of the European mass spectrometry community. The *Current IUPAC Recommendations* (see following sidebar) refers to the term *mass spectroscope* as essentially obsolete but uses it in the definition of mass spectrometry and then goes on to imply that the term *mass spectroscopy* is appropriate.

> **From the Latest IUPAC Recommendations**
>
> (considered to be obsolete and/or misleading by this author)
>
> *Mass spectrometry.* "The branch of science that deals with all aspects of mass spectroscopes and the results that are obtained with these instruments." *This definition is somewhat contradictory with respect to the following definition.*
>
> *Mass spectroscope.* "A term, which is now essentially obsolete, that refers to either a mass spectrometer or a mass spectrograph."
>
> *Mass spectroscopy.* "The study of systems that cause the formation of gaseous ions, which are characterized by their mass-to-charge ratios and relative abundances, with or without fragmentation." *This definition is somewhat contradictory to that of mass spectroscope.*
>
> *Mass spectrograph* and *mass spectrometer* are defined above.
>
> From the ambiguity of the above definitions, along with their contradictions, you have to wonder what the authors really mean.

Correct: **mass spectrometer** – a term first used by two well-known early mass spectrometry pioneers, William R. Smythe (U.S. scientist) and Josef Heinvich Elizabeth Mattauch (Austrian physicist), ca. 1926 (Kiser 1965) and is applied to those instruments that bring a focused beam of ions to a fixed collector, where the ion current is detected electrically. The term *mass spectrometer* is now used to describe all instruments that measure the abundance of ions based on their *m/z* values.

Correct: **mass spectrometry** – the study of matter based on the mass of molecules and on the mass of the pieces of the molecule. Mass spectrometry is often involved with mass spectra obtained with a mass spectrometer. The *Current IUPAC Recommendations'* definition of mass spectroscopy could more appropriately be applied to mass spectrometry than its definition of mass spectrometry.

Incorrect: **mass spectrograph** – when used to describe a specific type of *m/z* analyzer and should not be applied to a mass spectrometer. A mass spectrograph (first introduced by F. W. Aston in 1921) is an instrument that produces a focused mass (*m/z*) spectrum on a focal plane, where a photographic plate may be located. These instruments are capable of a high degree of mass accuracy but are not very good at determining ion abundances.

Correct: *m/z* **analyzer** or *m/z* **analysis** – terms that better describe the mass spectrometer and its functions than the terms *mass analyzer* or *mass analysis*. The use of *m/z* analyzer and *m/z* analysis is supported through the continual proliferation of mass spectrometers that use the electrospray technique that produce multiple-charge ions of high mass.

Incorrect: **mass spectrophotometer** – a term that never had any official recognition in mass spectrometry. There are no light bulbs in a mass spectrometer.

Correct: **mass spectrometrist** – the term used to describe a person who uses mass spectrometry. The term "mass spectroscopist" is **incorrect**.

Correct: **selected-ion flow tube mass spectrometry (SIFTMS)** – a technique used for the analysis of trace organic components in gas samples such as pollutants in air and metabolite gases in breath, using ionization that takes place in a fast-flow tube. H_3O^+, NO^+, and $O_2^{+\bullet}$ reagent ions are produced in a chemical ionization source using moist ambient air (p $\cong$ 50 Pa) as the reagent gas and a microwave gas-discharge ion source (as opposed to the ohmically heated filaments used to produce electrons in the conventional CI ion source) that provides an increased degree of ruggedness. The reagent (precursor) ions are selectively separated from other ions with a transmission quadrupole *m/z* analyzer. The reagent ions are injected into a **fast-flow tube** (diameter = 1–2 mm, length = 30–100 cm) that has He at temperature of ~300 °K following at a linear velocity of 40–80 m sec^{-1} maintained by using a Roots pump (p $\cong$ 100 Pa). Samples are introduced into the flow through a heated sampling line. Because the sample gas also contains water molecules, the reagent ion complex becomes more complicated with the presence of protonated water clusters and NO^+ water clusters, which along with the primary reagent ions can form adducts with analyte molecules. A second transmission quadrupole *m/z* analyzer is used to determine the amount of product ions formed and the reduction of precursor ions in the fast-flow tube. The second transmission quadrupole *m/z* analyzer is operated in full scan or selected-ion-monitoring mode. Quantitation is carried out through the use of a series of computer algorithms based on the physics of the ion/molecule reactions that take place in the fast-flow tube (Smith 2005). A variation of this technique that uses only the H_3O^+ is termed **proton-transfer reaction mass spectrometry (PTRMS)**.

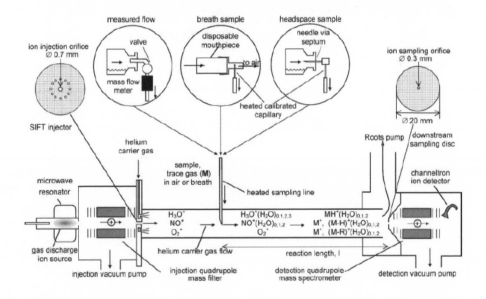

Figure 6. *Schematic representation of the selected-ion flow tube mass spectrometer.* (Smith, D; Španěl, P "Selected Ion Flow Tube Mass Spectrometry (SIFTMS) for On-Line Trace Gas Analysis" *Mass Spectrom. Rev.* **2005**, *24*, 661–700)

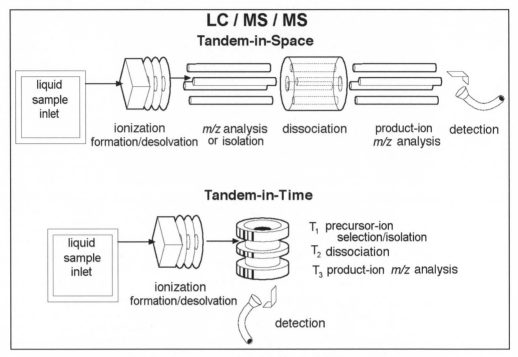

Figure 7. Comparison of tandem-in-space vs tandem-in-time.
(Johnson, JV; Yost, RA; Kelley, PE; Bradford, DC "Tandem-in-Space and Tandem-in-Time Mass Spectrometry: Triple Quadrupoles and Quadrupole Ion Traps" *Anal. Chem.* **1990**, *62*, 2162–2172)

Correct: **mass spectrometry/mass spectrometry (MS/MS)** (a.k.a. **tandem mass spectrometry**) – a mass spectral technique in which ions are caused to change mass (usually via decomposition but can form a heavier product from collision with a "reactive" neutral) to produce information that may not be obtainable from an initial ionization of the analyte. MS/MS is a technique where ion formation/fragmentation and subsequent decomposition of the original ions is carried out "in tandem". Using multiple *m/z* analyzers and an ion beam results in "tandem-in-space" (e.g., triple-quadrupole and quadrupole TOF mass spectrometers). Using ion trap mass spectrometers (quadrupole or magnetic) results in "tandem-in-time". In a tandem-in-space MS/MS instrument, there is a **collision cell** between the two *m/z* analyzers. This collision cell contains a **collision gas** at a pressure of 10^{-1} Pa. The collision gas is an inert gas that will result in inelastic collision with the precursor ion to bring about collisional activation. The collision gases most often used in the collision cell are He, Ar, Xe, and N_2. The He bath gas used to cool the ions in the quadrupole ion trap (QIT) is usually used as the collision gas. It is possible to replace the He in the QIT with one of the heavier collision gases such as Ar, Xe, or N_2 to bring about collisional activation of the precursor ion; and then replace the heavier collision gas with He to restore the mass spectral resolution before scanning the trap to obtain a product-ion spectrum.

Another secondary fragmentation technique that is analogous to MS/MS is postsource decay (PSD), which occurs in TOF instruments. PSD is defined in the **Terms Associated with Time-of-Flight Mass Spectrometers** section of this book.

There are three types of MS/MS analyses:

product-ion analysis (sometimes inappropriately referred to as **common-precursor-ion analysis**): This technique involves the isolation of a precursor ion, the collisionally activated dissociation of the precursor ion, and the production of a mass spectrum of the ions that result from this induced fragmentation. This type of analysis can be conducted in an instrument that involves the transportation of an ion beam from an area where the precursor ion is selected into the collision region followed by an area where the product ions are sorted, such as a: (1) triple-quadrupole mass spectrometer, (2) reverse-geometry double-focusing mass spectrometer, or (3) hybrid instrument composed of a quadrupole and a time-of-flight mass spectrometer. All of these instruments use two *m/z* analyzers separated by a collision cell. This analysis can also be carried out in a quadrupole ion trap or Fourier transform ion cyclotron resonance (FTICR) mass spectrometer. When carried out in one of these two latter types of instruments, a specific product ion is isolated by ejecting all other ions, and a dissociation is carried out. The dissociation process can be repeated with one of the product ions. The process can be carried out on several more generations of product ions. This process of generation of ions from the fragmentation of various generations of product ions is called **MS^n** where **n** is the number of ion formations.

precursor-ion analysis (sometimes inappropriately referred to as **common-product-ion analysis**): All of the ions formed by a primary ionization pass, one *m/z* value at a time, into the collision cell. The second *m/z* analyzer is set to allow only a single *m/z* value to reach the detector. This process results in the identification of only precursor ions that produce product ions of a specific *m/z* value. This type of MS/MS analysis can only be performed in instruments that use an ion beam.

common-neutral-loss analysis: All of the precursor ions formed by a primary ionization are allowed to pass, one *m/z* value at a time, into the collision cell. The second *m/z* analyzer is set to allow all ions to pass to the detector, one *m/z* value at a time. However, the *m/z* value of the second analyzer is offset by a fixed *m/z* difference from that of the first analyzer. This process results in the detection of analytes that have a common neutral loss. This type of MS/MS analysis can only be performed in instruments that use an ion beam.

It should be noted that these different "analyses" are often referred to as "scans", which is inappropriate even though the precursor-ion analysis and common-neutral-loss analysis can only be performed in scanning instruments that analyze an ion beam.

MS/MS can also be carried out in a hybrid instrument that uses an ion trap in conjunction with a beam-type *m/z* analyzer, thereby allowing for all three types of MS/MS analyses.

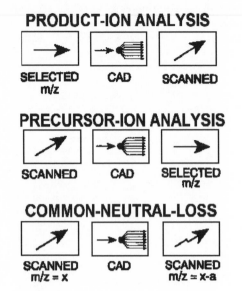

PRODUCT-ION ANALYSIS

SELECTED   CAD   SCANNED
m/z

PRECURSOR-ION ANALYSIS

SCANNED   CAD   SELECTED
m/z

COMMON-NEUTRAL-LOSS

SCANNED   CAD   SCANNED
m/z = x   m/z = x-a

Figure 8A. MS/MS symbolism proposed by E. de Hoffmann.
(Adapted from de Hoffmann, E "Tandem Mass Spectrometry: A Primer" *J. Mass Spectrom.* **1996**, *31*, 129–137)

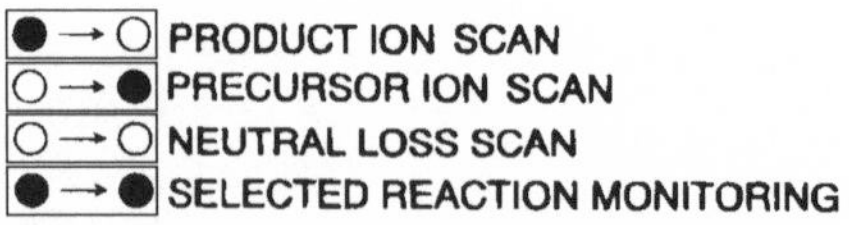

● → ○ PRODUCT ION SCAN
○ → ● PRECURSOR ION SCAN
○ → ○ NEUTRAL LOSS SCAN
● → ● SELECTED REACTION MONITORING

Figure 8B. Symbolism proposed by Kondrat and Cooks to represent the various scan modes; ● *refers to a fixed mass analyzer, and* ○ *refers to a scanning [sic] mass spectrometer.*
(Kondrat, RW; Cooks, RG *Anal. Chem.* **1978**, *50*, 81A)

When a **precursor ion** of a specific *m/z* value dissociates due to collisional activation and produces a **product ion** of a specific *m/z* value, the term **transition** is used (e.g., "the transition from *m/z* 379 to *m/z* 301" or "a 78-Da transition was observed").

It should be noted that when multiple-charge ions are caused to fragment, the fragments can have a higher *m/z* value because the number of charges can be reduced to a greater extent than the mass of the ion.

The MS/MS acronym was first coined for "mass spectrometry/mass spectrometry" by William F. Haddon in a symposium organized by Michael L. Gross, presented at the University of Nebraska, Lincoln, Nebraska, November 3–5, 1976 (Haddon 1978), to describe the technique of bringing about a decomposition of a stable ion by a forced collision with neutral gas molecules. The acronym was later defined by F. W. McLafferty and F. M. Bockhoff as "mass separation/mass spectral characterization" (MS/MS) by analogy to gas chromatography/mass spectrometry (McLafferty 1978); however, mass spectrometry/mass spectrometry is the definition of MS/MS that is used today. The symbolism proposed by E. de Hoffmann (de Hoffmann 1996) for the three different types of MS/MS analyses and that of R. W. Kondrat and R. G. Cooks (Kondrat, Cooks 1978) are shown in **Figures 8A** and **8B**, respectively.

Collisional Activation Conditions for Different Systems

Instrument Type	Collision Gas	Number Collisions	Collision Energy (eV)	Collision Time (µs)
Nontrap Hybrids, (Triple Quad, Q-TOF)	Ar, Xe, N₂	Few	10–100	5–50
Double Focusing & TOF-TOF	He	Single	2000–4000	<5
All QIT Type	He	Multiple	6–10	20,000–40,000
Insource	N₂	Few	1–400	5–50

Correct: **insource CAD** – similar to MS/MS in that the results are product ions produced by a collision of precursor ions. Insource CAD is used with the API techniques of APCI and ES. There is no *m/z* selection of a precursor ion by mass spectrometry. The only selection is accomplished through chromatographic purification of the precursor molecule. The precursor ions are fragmented in the moderate pressure area of the ion source because they undergo activating collisions with the gas (in most cases, nitrogen molecules) that is present. The fragmentation efficiency is based on a controllable velocity of the ions as they enter the *m/z* analyzer. This technique requires a limit to the number of different *m/z* values of precursor ions. Insource CAD has also been referred to as **transport-region CID**.

Correct: **surface-induced dissociation** (**SID**) – the fragmentation of a precursor ion brought about by a collision with a solid surface. SID has the advantage of not requiring a relatively high-pressure collision gas, and the efficiency approaches 100% when large energy transfers occur.

Correct: **selected reaction monitoring** (**SRM**) – an MS/MS technique that is similar to selected-ion monitoring (SIM) but may allow for a much higher degree of specificity. In SRM, ions of a specific *m/z* value are allowed to pass into the collision cell. The second *m/z* analyzer is set to allow only product ions of a specific *m/z* value to pass to the detector (i.e., the precursor ion must undergo a selected transformation or reaction to result in a response from the detector). SRM applies regardless of whether there is more than one precursor ion and multiple product ions from a single precursor ion, just as SIM applies when there is more than one ion monitored.

Incorrect: **multiple reaction monitoring** (**MRM**) – when used as a synonym for SRM. This neologism was created by an instrument manufacturer to distinguish their instruments from instruments of other manufacturers. This term is misleading as to what is actually being monitored in the mass spectrometer and should not be used. In the case where the SRM analysis involves multiple generations of product ions (e.g., ion trap mass spectrometers), MRM has been used to show that more than one generation of product ions are being monitored, which gives MRM two different meanings and leads to even more confusion.

Note: In *MS/MS analyses*, ions undergo an inelastic collision with neutral molecules in a collision cell; and in *insource CAD*, ions collide with neutral molecules in the ion source. When these collisions take place, a number of different things can happen to the precursor ion. The most common is collisionally activated dissociation.

$$m_p^+ \; + \; N \longrightarrow m_d^+ \; + \; m_n \; + \; N$$

However, the following three reactions can also take place. Because of the nature of API techniques, these reactions can be significant in insource CAD.

charge exchange:

$$m_p^{+\bullet} \; + \; N \longrightarrow m_p \; + \; N^{+\bullet}$$

partial charge transfer:

$$m_p^{2+} \; + \; N \longrightarrow m_p^{+\bullet} \; + \; N^{+\bullet}$$

charge stripping: This process is an ion/molecule reaction that increases the number of positive charges on an ion. Charge stripping is an ionization process.

$$m_p^{-\bullet} \; + \; N \longrightarrow m_p^{+\bullet} \; + \; N \; + 2e^-$$

$$m_p^{+\bullet} \; + \; N \longrightarrow m_p^{2+\bullet} \; + \; N \; + \; e^-$$

$$m_p \; + \; N \longrightarrow m_p^{+\bullet} \; + \; N \; + \; e^-$$

It is possible that charge stripping can be combined with a mass change.

$$m_p^- \; + \; N \longrightarrow m_d^+ \; + \; (m_p - m_d) \; + \; N + 2e^-$$

If the collision gas is a chemically active substance, then an association reaction can take place to yield an adduct ion (or even a condensation product) that has a mass that is greater than the precursor ion.

$$m_p^+ \; + \; m_n \longrightarrow (m_p^+ + m_n)$$

Another ion/neutral reaction is **charge inversion**. This process is where the sign of the charge is reversed. The first example of charge stripping, above, is also charge inversion.

All of these ion/neutral reactions are **charge permutations**. This general term describes an ion/neutral species reaction where a change occurs in the magnitude and/or sign of the charge (de Hoffmann 1996).

Correct: **peak matching** – a technique used with double-focusing instruments operating in the high-resolution mode (typically **M/ΔM** $\geq$10,000) where a particular monoisotopic peak of an unknown is superimposed to a peak of an internal standard (the exact mass of which can be calculated); the two molecules, unknown and internal standard, must be different but close in mass. The magnet position is fixed, and the accelerating voltage is switched between a known peak and a reference peak. The accelerating voltage difference can be used to calculate the mass difference between the unknown peak and the standard peak. The peak matching must first be calibrated using two reference points to define the slope and intercept, which is done prior to making a series of measurements. A second magnet coil is sometimes used, but its purpose is just to sweep the peak over a narrow range so that you can see the peak shape on the oscilloscope. The mass measurement comes from the accelerating voltage (and electric sector) switching. In some cases, peak matching is done using a single peak of known *m/z* value as an internal standard. The ion represented by this peak is referred to as the **lock mass**.

Correct: **secondary-ion mass spectrometry (SIMS)** – a technique that uses the bombardment of a solid sample with a high-energy beam of ions to produce a mass spectrum of secondary ions that are generated on the surface of the sample. This technique was originally developed for the analysis of inorganic materials and polymeric surfaces. *Static* SIMS uses lower primary-ion dose densities at pressures of 10^{-8} Pa. This technique is employed for surface analysis in the *x–y* plane and is used in organic applications. *Dynamic* SIMS uses a higher ion flux to achieve depth-profile measurements along the *z* axis of the sample.

Correct: **selected-ion monitoring (SIM)** – a chromatographic/mass spectrometric technique where ion current at only one or a few selected *m/z* values is detected and stored during the chromatographic separation. This is in contrast to the other mass spectral data acquisition technique used with chromatography, **continuous measurement of spectra (CMS)**, a process where spectra are acquired over a specified *m/z* range, one after another (Budde 2001). According to J. T. Watson (Watson 1997), "the term *selected* is appropriate because it implies both choice and specificity; furthermore, it imposes no restriction as to the number of ions involved. The word *ion* accurately describes the species being monitored. The term *monitoring* is preferred because it connotes the element of time in this specialized technique which records profiles of ion current as a function of time." Either the full phrase or the abbreviation can be used in print. When spoken, each letter should be pronounced, and the acronym should not be used. It is easy to confuse SIM with SIMS, which is spoken as the acronym for secondary-ion mass spectrometry.

Incorrect: **selective ion monitoring** – when used as a phrase that indicates the ions are doing the selecting of what to monitor, not that they were selected by the analyst for observation. This phrase has appeared occasionally in print but is incorrect. The term **mass-selective detector (MSD)** is correct because it refers to the fact that the detector monitors only ion currents at certain "masses" (*m/z* values where *z* is always 1).

Note: **multiple-ion detection (MID)** – a product, originally sold by LKB Instruments, that performed an SIM analysis. Finnigan uses this term to describe the monitoring of the ion current of a few specific *m/z* values. This term is considered to be archaic and should not be used.

The first term used to describe the technique of selected-ion monitoring in magnetic-sector instruments is **accelerating voltage alternation (AVA)** (Sweeley 1966). AVA is highly specific to the instrument type and is no longer considered to be correct. A third term used to describe the SIM technique is **mass fragmentography** (Hammer 1968). Mass fragmentography is considered to be archaic and should not be used.

SIM has also been used as an abbreviation for the term *single-ion monitoring*, another name used for the technique. Single-ion monitoring is too restrictive and should not be used. The term **selected ion** recording (**SIR**) has been used by Micromass.   With the latest Micromass addition, according to the 13 different terms listed and referenced by J. T. Watson (Watson 1974), SIM has had at least 14 different names:

accelerating voltage alternation	selected-ion monitoring
mass fragmentography	multiple-ion monitoring
multiple-ion detection	multiple mass monitoring
multiple-specific-ion detection	multiple-ion analysis
selective ion detection	tuned-ion analysis
ion-specific detection	selected-ion-peak recording
selective ion monitoring	selected ion recording

In another interesting Letter-to-the-Editor of *Biomedical Mass Spectrometry* by Fred Falkner, a former postdoc with J. T. Watson at Vanderbilt University (Falkner 1977), it is reported, "The earliest publication of the technique of using a mass spectrometer with a single detector to detect, as a function of time, the ion current as a few selected masses appears to be that of Nier et al. in 1948." The first application of this technique to organic GC/MS is credited to Sweeley et al. in 1966.

Questionable: **selected ion storage** (**SIS**) – a term unique to the Varian Saturn quadrupole ion trap GC-MS by which the trap is filled only with ions of desired *m/z* values by ejection ions of all other *m/z* values prior to ion detection. In the case of samples that have high amounts of coeluting matrix components, SIS will give a decreased detection limit.

Proposed: **selected-ion-summation analysis** (**SIS analysis**) – a data analysis technique proposed in 1977 by D. W. Kuehl (Kuehl 1977) by which ion intensity profiles produced by postdata acquisition from a reparative scanning GC/MS analysis are summed and reported only if all the individually summed ions exhibit abundances in the same chromatographic peak. This technique never became popular due to the lack of commercial software.

Note: Although there is not much chance of confusion, the SIS acronym should not be used without first defining it.   Unlike SIM and SRM, the SIS acronym has not been assumed.

Correct: **homolytic** (a.k.a. **radical-site-driven**) **cleavage** – a fragmentation that results from one of a pair of electrons between two atoms moving to form a pair with the odd electron. After fragmentation, the atom that contains the charge when the ion is formed retains the charge. A radical is lost as a result of the fragmentation. This reaction involves the movement of a single electron and is symbolized by a single-barbed arrow, the so-called "fishhook" convention (Budzikiewicz 1964).

Unfortunately, the *Current IUPAC Recommendations* states that the symbol that indicates the movement of one electron (homolysis) and the symbol for the movement of two electrons (heterolysis) is the same, a double-barbed arrow.

*Correct: **alpha** (**α**) **cleavage** (**a special form of homolytic cleavage**) – a fragmentation (homolytic cleavage) that results from one of the pair of electrons between the atom attached to the atom with the odd electron and an adjacent atom that pairs with the odd electron. After fragmentation, the atom that contains the charge when the ion is formed retains the charge. A radical is lost as a result of the fragmentation. This fragmentation is homolytic cleavage because it involves the movement of a single electron (McLafferty 1973).

$$R_1 - \overset{\overset{\displaystyle \overset{+}{\cdot \cdot}}{\|}}{C} - \overset{c}{\underset{}{CH_2}} - R_2 \longrightarrow R_1 - C \equiv \overset{+}{\underset{}{O}} \cdot : \\ + \\ \cdot CH_2 - R_2$$

*Incorrect: **alpha (α) cleavage** – when defined as "...fission of a bond originating at an atom that is adjacent to the one assumed to bear the charge; the definition of β, γ, [etc.] then follows automatically" (Budzikiewicz 1964). This definition allows the use of α-cleavage to describe a bond fission that results in original charge-site retention (homolytic cleavage) or charge-site migration as a result of bond fission (heterolytic cleavage). Using α cleavage with this somewhat ambiguous definition can lead to confusion. The convention established by McLafferty (McLafferty 1973) in which the term "α-cleavage" is used to define a special case of homolytic fission results in a clearer communication. In an attempt to reduce the confusion created from the use of this term, some authors use "α-cleavage with charge retention" and "α-cleavage with charge migration". The *Current IUPAC Recommendations* uses the Budzikiewicz, Djerassi, and Williams recommendation as the definition of α-cleavage (Budzikiewicz 1964). See **Abbreviation Usage** section.

Correct: **benzylic cleavage** – a fragmentation that takes place at the carbon atom attached to a phenyl group. When the phenyl group is C_6H_5, the benzylic cleavage will result in a benzyl ion with a formula of $C_6H_5=CH_2^+$, which can be isomeric with the tropylium ion. Benzylic cleavage is a special case of homolytic cleavage and is due to the loss of a pi (π) electron, which places the site of the charge and the radical on the phenyl ring.

Correct: **heterolytic** (a.k.a. **charge-site-driven** or **inductive**: **i**) **cleavage** – a fragmentation that results from the pair of electrons between the atom attached to the atom with the charge and an adjacent atom that moves to the site of the charge. This fragmentation involves the movement of the charge site to the adjacent atom. A radical is lost as a result of the fragmentation. The movement of a pair of electrons is symbolized by a double-barbed arrow (Budzikiewicz 1964).

Following is an example of the fragmentation of a heterolytic cleavage.

OR

Correct: **γ-hydrogen shift-induced beta (β) cleavage** (a.k.a. the **McLafferty rearrangement**) – a rearrangement reaction that was originally described by an Australian chemist, A. J. C. Nicholson (Nicholson 1954), but was named after F. W. McLafferty (McLafferty 1993) because of the extent to which he studied and reported the reaction in a wide variety of compound types. An odd-electron fragment ion is formed by the loss of a molecule. This fragment results from a gamma-hydrogen (γ-hydrogen) shift to an unsaturated group such as a carbonyl (when the site of the odd electron and the charge is on the oxygen atom). The γ-hydrogen shift causes the radical site to move to the carbon atom that originally contained the γ hydrogen. This new location of the radical site initiates a cleavage reaction that causes the fragmentation of the carbon–carbon bond that is beta to the unsaturated group and the loss of a terminal olefin.

m/z = 298   m/z =74

Correct: **blackbody infrared radiative dissociation (BIRD)** – dissociation of an ion induced by the absorption of infrared photons radiated from a heated blackbody, which is usually the walls of a vacuum chamber. BIRD is a special case of infrared multiphoton dissociation (IRMPD). This technique is most often used in FTICR mass spectrometers (Dunbar 2004).

Correct: **infrared multiphoton dissociation (IRMPD)** – a process of energizing ions to induce fragmentation by using adsorption of multiple infrared photons. This technique is usually carried out by FTICR mass spectrometers using a continuous wave of 10.6-μm CO_2 laser with a 25–40-W power passed into the ICR cell through a ZnSe or BaF_2 window. Irradiation is typically varied between 5–300 ms (Laskin 2005).

Correct: **charge-remote fragmentation** – occurs in straight-chain ions of the form $H_3C-(CH_2)_m-CH_2-CH_2-CH_2-CH_2-(CH_2)_n-X^+$ where the site of the charge is on X. This is prevalent in high-energy CAD; and, to a lesser extent, in low-energy CAD. The products are $H_3C-(CH_2)_m-CH=CH_2$, $H_2C=CH-(CH_2)_n-X^+$, and H_2 via a 1,4-elimination. A good example of this type of fragment is found in spectra produced by high-energy CAD using the protonated molecule of a fatty acid as the precursor ion.

Correct: **charge-site derivatization** – a process of forming a derivative that will ensure the location of the charge site in an ion so as to force the fragmentation to produce a specific ion series. The derivatization covalently bonds a specific functional group with a moiety that has a higher proton affinity than any possible charge site on the underivatized molecule. This is often employed in the CAD analysis of peptides in order to produce a specific ion series.

Correct: **hydrogen shift** – the movement of a hydrogen atom, usually in response to a radical site. A hydrogen shift results in a hydrogen proton and one of the two electrons in the bond between the hydrogen and an adjacent atom moving away from the ion to another location on the ion. A bond is broken, and a bond is formed.

Correct: **hydride shift** – the movement of a hydrogen proton with the two electrons that attach it to an adjacent atom. This movement results in a bond being broken and a new bond being formed.

A hydride shift

Correct: **ion pair** – a neutral species that is formed from a positive ion and a negative ion that each have a single charge. Ion pairs can also be formed from more than a single positive ion and a single negative ion; however, the net charge must be zero. The term **ion-pair formation** means that both a positive and a negative ion are formed in a fragmentation.

Correct: **EIEIO** – a term associated with the farm or the animals on the farm of a man named MacDonald, affectionately referred to as "Old MacDonald". **EIEIO** has also been used in mass spectrometry as an acronym for **electron-induced excitation in organics** (McLafferty 1993) and **electron impact excitation of ions from organics** (Cody 1979). This technique involves the excitation of trapped ions with a continuous electron beam. The spectra obtained by EIEIO in an ion cyclotron resonance mass spectrometer are analogous to spectra that result from CAD and yield characteristic structural information. The technique is also known as **electron-induced dissociation**.

Correct: **electron capture dissociation** (**ECD**) – a technique used to induce the dissociation of multiple protonated species through the capture of low-energy (<0.2 eV) electrons. This is another technique that is employed with FTICR mass spectrometers. The resulting odd-electron ion ($[M + nH]^{n+} + e^- \rightarrow [M + nH]^{(n-1)+\,\bullet}$) will readily fragment. This technique is very useful in the study of peptides because posttranslational modifications (PTMs) such as phosphorylation, O– and N–linked glycosylation, and sulfation are preserved, allowing for site-specific analyses (Cooper 2005).

Correct: **even-electron rule** – odd-electron ions (e.g., molecular ions and fragment ions resulting from rearrangements that eliminate a molecule, an even-electron neutral) can fragment with the loss of either a molecule (an even-electron species) or a radical (an odd-electron species); however, even-electron ions (e.g., protonated molecules, $[M - 1]^{+\ or\ -}$, produced by either hydride or proton abstraction, respectively, or fragment ions produced by single-bond cleavage of positive-charge molecular ions produced by electron ionization) do not lose a radical to form an odd-electron ion. The tendency is for even-electron ions to fragment with the loss of an even-electron molecule (Karni 1980).

$$[\text{odd}]^{+\bullet} \longrightarrow [\text{even}]^+ + R^\bullet$$
$$[\text{odd}]^{+\bullet} \longrightarrow [\text{odd}]^{+\bullet} + M$$
$$[\text{even}]^+ \longrightarrow [\text{even}]^+ + M$$
$$[\text{even}]^+ -\!/\!\!/\!\rightarrow [\text{odd}]^{+\bullet} + R^\bullet$$

Correct: **Field's rule** – associated with the fragmentation of even-electron ions. The tendency for a neutral fragment to leave depends on its proton affinity (PA). The formation of $C_2H_5^+$ is greater from $C_2H_5O^+{=}CH_2$ (through charge migration) than from $C_2H_5S^+{=}CH_2$ because the PA of $O{=}CH_2$ is less than that of $S{=}CH_2$ (7.4 eV vs 8.9 eV). The lower the PA of the neutral molecule, the greater the tendency for it to leave the even-electron ion (Field 1972).

Correct: **octet rule** – the statement that no energy shell of an atom can hold more than eight electrons, as long as it is the outermost shell of the atom.

Correct: **ortho effect** – a term associated with the effect of a group with a labile hydrogen adjacent to a group that can carry the site of the charge in an electron ionization on an aromatic ring. This adjacent relationship causes a hydrogen rearrangement, resulting in the loss of a molecule by charge migration to produce an odd-electron fragment ion. This behavior is not seen when the two groups are *para* or *meta* to one another. Examples are seen in the mass spectra of *o*-chlorophenyl and salicylic acid.

Correct: **pi (π) bond** – a bond formed by the side-by-side overlap of the *p* suborbitals of the outer energy shells of two adjacent atoms. A double bond will have one pi bond and one sigma bond; a triple bond will have two pi bonds and one sigma bond.

Correct: **rearrangement** – a number of different fragmentations occur as a result of a rearrangement. This usually involves the breaking of two bonds and the formation of a third bond. Examples of rearrangements are the γ-hydrogen shift-induced β cleavage and the hydride shift. Some rearrangements have been observed that involve a hydrogen and a hydride shift due to the presence of a tautomerism of double bonds that can take place when two heteroatoms with nonbonding electrons are involved. There are other examples of these triple-bond fragmentation rearrangements that always involve a distonic ion.

Correct: **retro-Diels–Alder reaction** – a cleavage that results from the breaking of two bonds to form a butadiene odd-electron ion and a neutral even-electron olefinic fragment (a molecule) or an odd-electron product ion by the explosion of a molecule of a butadiene. This reaction is the reverse of the 1,4-addition of an olefinic unit to a conjugated diene (a Diels–Alder reaction). The retro-Diels–Alder fragmentation is often found in the mass spectra of cyclic olefins (Biemann 1962).

Correct: **sigma (σ) bond** – a bond formed by the end-to-end overlap of the *sp* hybridized suborbitals of the outer energy shell of two adjacent atoms or the overlap of an *sp* hybridized suborbital of the outer shell of an atom and the *s* suborbital of the single energy shell of the hydrogen atom. The σ bond can be formed using *sp*, sp^2, or sp^3 hybrid orbitals.

Correct: **sigma-bond (σ-bond) cleavage** – a fragmentation of a molecular ion that is formed by the loss of a σ-bond electron during the initial ionization. A radical is lost as a result of the fragmentation.

Correct: **skeletal rearrangement** – a rearrangement that involves the shift of a hydrogen atom or a hydride ion that results in a new radical site or charge site, respectively. The McLafferty rearrangement is an example of a skeletal rearrangement, as is the loss of water from a protonated alkyl aldehyde or the periodicity of peaks that differ by 56 *m/z* units in the mass spectrum of methyl stearate. The same symbolism used to indicate the McLafferty rearrangement is used to indicate a skeletal rearrangement. A good example of skeletal rearrangements is shown in **Figure 9**, which is adapted from the works of Ryhage, Sonneveld, and Stenhagen.

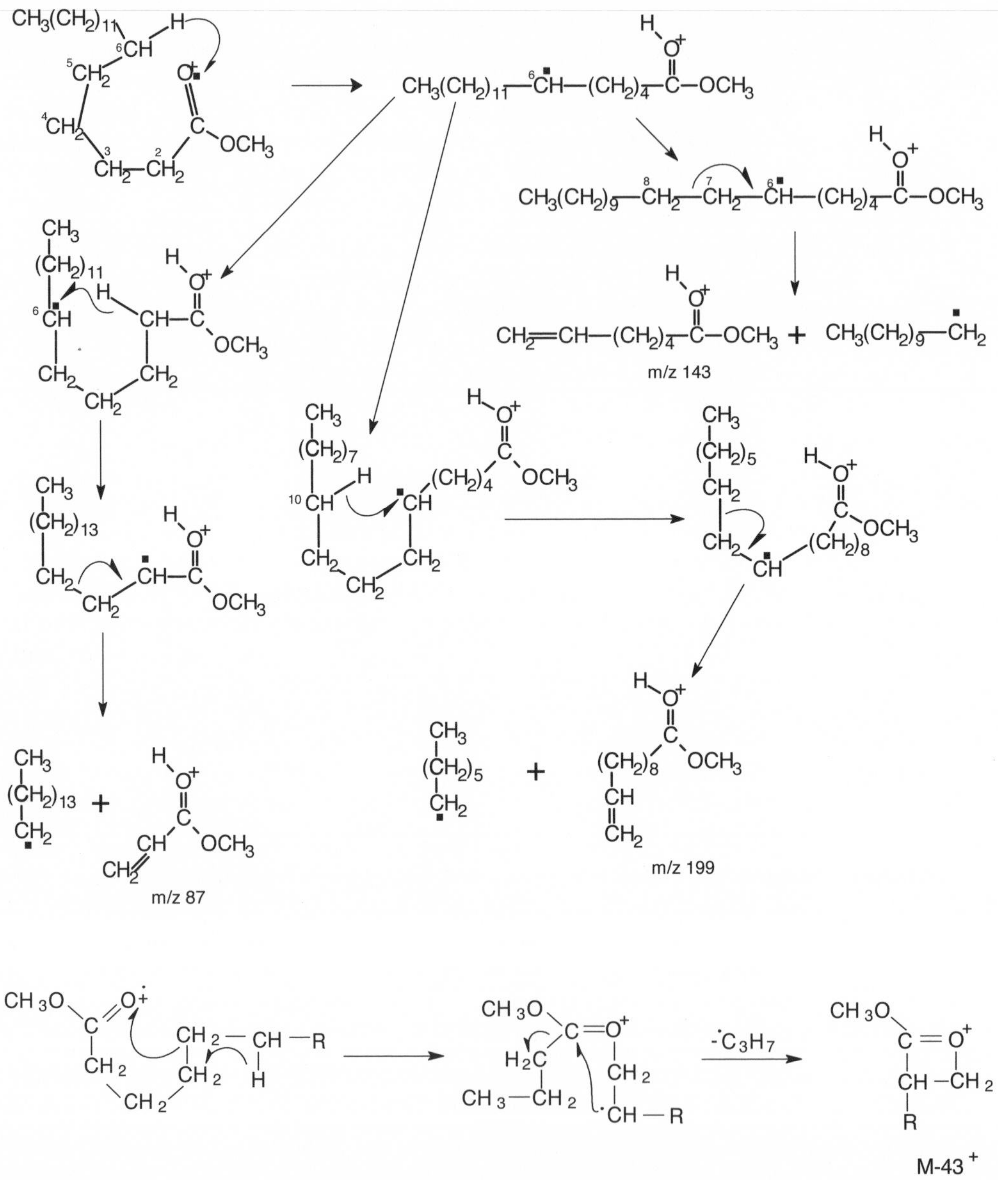

Figure 9. Use of skeletal rearrangements to explain peaks of anomalous intensities in the electron ionization mass spectrum of methyl stearate.

Adapted from: (1) Ryhage, R; Stenhagen, E Mass Spectrometry of Long-chain Esters. In McLafferty FW, Ed., *Mass Spectrometry of Organic Ions*; Academic: New York, 1963, 399–452; (2) Stenhagen, E "Current State of Mass Spectrometry for Organic Analysis" *Z. Anal. Chem.* **1964**, *205*, 109–124; (3) Budzikiewicz, Z; Djerassi, C; Williams, DH *Interpretation of Mass Spectra of Organic Compounds*; Holden-Day: San Francisco, CA, 1964; p 15; (4) Sonneveld, W, Ph.D., Thesis, Delft, 1967.

Correct: **Stevenson's rule** – associated with sigma-bond (σ-bond) cleavage. The σ-bond cleavage of an odd-electron ion leads to two sets of ion-radical products: $ABCD^{+\bullet}$ will produce A^+ and $^\bullet BCD$, or DCB^+ and $^\bullet A$. The radical that has the highest tendency to retain the odd electron will also have the higher ionization energy. Therefore, the fragment with the lowest ionization energy will be preferentially formed. This lower-energy ion should be more stable; therefore, the more abundant (Stevenson 1951). A notable exception to Stevenson's rule is the preference for the **loss of the largest alkyl** radical at the site of ionization. In a series of secondary carbenium ions produced from an aliphatic hydrocarbon molecular ion that has a methyl, ethyl, and butyl group along with a hydrogen atom attached to a carbon, the most stable ion would be the one that results from the loss of the hydrogen radical; however, it will be the least abundant ion. The most abundant will be the ion produced by the loss of the butyl radical, which is the largest of the four possible losses. This ion is also the least stable.

Inappropriate: **simple cleavage** – when used to describe the unimolecular fragmentation of single-charge molecular ions ($M^{+\bullet}$) to produce an even-electron fragment ion (EE^+) and an odd-electron radical ($OE^\bullet$) by the breaking of a single bond. This term was first introduced by McLafferty (McLafferty 1973). It is an all-encompassing term that can refer to σ-bond, homolytic, or heterolytic cleavage. With such an overall ambiguous term, its use could result in rampant confusion. Another term (resulting in less confusion) often used in conjunction with simple cleavage is **dissociation with rearrangement**, which is defined as the fragmentation of a $M^{+\bullet}$ through a process in which bonds are broken and new bonds are formed.

Correct: **multiple-bond cleavage** – a term that describes a process of ion fragmentation resulting in the breaking of more than one bond. Multiple-bond cleavage is associated with a rearrangement.

Correct: **single-bond cleavage** – a term that describes the loss of a radical from a molecular ion via σ-bond, homolytic, or heterolytic cleavage.

Correct: **unimolecular** – a term used in mass spectrometry in relation to the decomposition of an ion. The term has its origin in EI mass spectrometry. "(EI) mass spectral reactions are *unimolecular*; the sample pressure in the ion source is kept sufficiently low that bimolecular ("ion-molecule" [sic]) or other collision reactions are usually negligible" (McLafferty 1973). Decompositions of ions formed through ion/molecule reactions as a result of collisional activation also involve only a single species.

A Philosophical Discussion

When discussing mass spectral data, the issue of fragmentation often comes to the forefront. Many terms used in a discussion of fragmentation are defined in this section. Some terms have been omitted because they are self-explanatory, or they are included in the definition of other terms. The terms **charge-site-initiated fragmentation** and **radical-site-initiated fragmentation** are implicitly self-explanatory. These terms are also incorporated to some extent in the definitions for **heterolytic** and **homolytic cleavages**, respectively.

One term often used in the discussion of mass spectral data is **diagnostic ion**. This term is usually used in reference to a peak in a mass spectrum that represents a diagnostic ion such as a peak representing a phenyl ion (a peak at *m/z* 77), indicating the presence of a phenyl moiety on a precursor ion.

Another term used in the discussion of a mass spectrum is **ion series**. This term usually means a series of peaks in an EI mass spectrum that indicates structural moieties such as a series of peaks spaced 14 *m/z* units apart, indicating the presence of an alkyl moiety. Each of the peaks represents carbenium ions that differ in structure by a single –CH_2– group. Another example of an ion series that has a specific meaning could be peaks representing an ion that is 22 *m/z* units higher than the lowest *m/z* value peak in the series and a peak 17 *m/z* units lower than the peak at the highest *m/z* value in the series in a mass spectrum obtained using electrospray ionization. This series could indicate the presence of a peak representing a protonated molecule followed by a peak representing a sodiated molecule at the next higher *m/z* value followed by a peak representing a potassiated molecule at the highest *m/z* value in the series.

As has often been stated, "Ions are found in mass spectrometers, and peaks are a part of the recording of a mass spectrum." Always keep this in mind when presenting information obtained using mass spectrometry.

IONS

Correct: **acylium ion** – an even-electron ion that is the product of a single-bond cleavage, usually alpha (α) cleavage, of an odd-electron ion that contains oxygen, and the original site of the charge is on the oxygen atom. The acylium ion (usually referred to in EI mass spectrometry) has the form: $R\text{–}C{\equiv}O^+$ where R is an aliphatic; if R is a phenyl or substituted phenyl, then the ion is a **benzylium ion**.

Correct: **adduct ion** (a.k.a **adduct**) – an ion that results from an ion/molecule reaction where the ion attaches itself to the molecule. A *protonated molecule* is a special case of an adduct ion. Ions formed in the gas phase in chemical ionization such as $RC_2H_5^+$ and $RC_3H_5^+$ or gas-phase ions such as RNa^+, RNH_4^+, RCl^-, (etc.) produced in electrospray are adduct ions.

Correct: **allyl ion** – an even-electron ion that is the product of α cleavage initiated by ionization at an olefinic double bond (or phenyl π-electron system: **benzyl ion**). Allyl ions have the form: $CH_2{=}CH\text{–}C^+H_2 \leftrightarrow {}^+CH_2\text{–}CH{=}CH_2$.

Correct: **alkyl ion** – an even-electron ion that is the product of a single-bond cleavage, usually sigma-bond (σ-bond) cleavage, of a hydrocarbon ion (odd- or even-electron ion) that contains no aromatic groups.

Correct: **appearance energy** – the minimum energy that must be absorbed by an atom or molecule to produce a specified ion. The specified ion does not have to be a fragment ion. The appearance energy is less than 20 eV for most organic compounds.

Incorrect: **appearance potential** – when used as a term to specify the energy of the electrons in electron ionization at which fragment ions begin to appear. This term is no longer recommended by IUPAC and also because of the potential confusion with *appearance energy*.

Correct: **benzyl ion** – an even-electron ion that is the product of α cleavage of an odd-electron ion that is initiated by the radical and charge on the ring. The benzyl ion has the form: $C_6H_5\text{–}C^+H_2 \leftrightarrow H_2C{=}C_6^+H_5$. Unlike the tropylium ion (which has all 7 carbon atoms in the ring), the benzyl ion has only 6 carbon atoms in the ring.

Correct: **carbenium ion** – an even-electron hypovalent ion with the positive charge on a carbon atom usually formed by σ-bond cleavage. The three types of carbenium ions are shown below where R, R′, and R″ are generally any organic substructure. The order of stability is tertiary > secondary > primary. The **-enium** suffix can be used with a noncarbon positive-charge hypovalent ion such as an oxenium ion (RO^+).

$$R\text{—}\overset{+}{C}H_2 \qquad R\text{—}\overset{+}{C}H\text{—}R' \qquad R\text{—}\underset{R'}{\overset{R''}{\overset{|}{\underset{|}{C}}}}{+}$$

Primary   Secondary   Tertiary

Incorrect: **carbonium ion** – when used to describe a trivalent carbon ion. A *carbonium ion* is an even-electron hypervalent ion formed by the addition at a carbon site of a molecule to form a fifth covalent bond. An example of a carbonium ion is the primary CI reagent ion of methane, CH_5^+ (methonium ion). For many years, the positive even-electron ion of carbon with three subgroups (which is now known as a *carbenium ion*) was called a carbonium ion. Although the existence of the CH_5^+ ion in the gas phase had first been reported much earlier (Tal'rose 1952), when George Olah began to describe its existence in organic solution chemistry (Olah 1971, 1972), he proposed the change to the current nomenclature of carbonium and carbenium based on the evidence of an intermediate in which the positive charge at the carbon is a result of five covalent bonds. At the same time, it was proposed that carbonium and carbenium ions be referred to as *carbocations*. This nomenclature has been accepted by IUPAC (Gold 1983). The term *carbonium ion* sometimes (**incorrectly**) appears in current literature to describe the trivalent *carbenium ion*. The **-onium** suffix is also used with other hypervalent positive-charge ions (e.g., the oxonium ion RO^+H_2).

Incorrect: **carbocation** or **carbanion** – when used to refer to an organic positive or negative ion. These terms are incorrect because the use of *cation* as a synonym for a positive ion and the use of *anion* as a synonym for a negative ion is no longer recommended by IUPAC for use in mass spectrometry (see **positive ion** and **negative ion**).

Correct: **gas-phase cation** – an ion that is desorbed from solution into the gas phase that is not the result of a chemical reaction in the condensed phase. Examples of gas-phase cations would be quatenary amminium and phosphonium ions. The term **cation** should not be used in describing gas-phase ions such as positive molecular ions, protonated molecules (from solution or the gas phase), or positive adduct ions like sodiated molecules (see **positive ion** and **negative ion**).

Correct: **cluster ion** – a cluster is a species resulting from two or more neutral particles (usually molecules) that are bound together through noncovalent forces (e.g., H_2O clusters, analyte multimers, analyte molecules combined with molecules of mobile phase [e.g., Analyte + MeOH, Analyte + AcCN, etc.] etc.). A cluster ion is formed by the combination of an ion with one or more of another ion, atom, or molecule of a chemical species (e.g., $[(H_2O)_nH]^+$ is a cluster ion).

Correct: **deprotonated molecule** – an ion formed as a result of a gas-phase ion/molecule reaction or solution chemistry (ES) that results in the deprotonation of an analyte molecule to produce an even-electron negative ion, $[M - 1]^-$ or $[M - H]^-$.

Correct: **distonic ion (distonic radical ion)** – an odd-electron ion in which the radical and charged sites are separated (Radom 1984). The site of the charge and the radical site are associated with nonadjacent atoms. There is one exception to this definition, which is the α-distonic ion (a.k.a. **ylideion**, pronounced ilideion (Yates 1984), resulting from an ylide, an internal salt $[R_2^-C–P^+Ph_3]$ with a heteroatom formerly in a cation state) (e.g., $^\cdot CH_2OH_2^+$, $^\cdot CH_2ClH^+$, $R_2^\cdot C–N^+H_2R'$, etc.). There are also β-distonic ions (charge and radical site separated by one non-hydrogen atom), γ-distonic ions (charge and radical site separated by two non-hydrogen atoms), and so forth. Distonic ions result from rearrangements, usually hydrogen rearrangements.

Distonic Ion

Correct: **electron volt (eV)** – a unit of energy that is the work done on an electron when passing through a potential rise of 1 volt. 1 eV = 1.602×10^{-19} joules. The energy of the electron beam in electron ionization mass spectrometry is expressed in eV. In modern instruments, the ionization energy standard for EI mass spectrometry is 50–100 eV (i.e., electrons are accelerated between 50 and 100 volts in the ion source). In the early days of mass spectrometry, the standard was set at 70 eV. One electron volt is equivalent to 23 kcal of energy.

Correct: **electron energy** – the numerical description that results from the potential difference through which electrons are accelerated in electron ionization. The term **ionization energy** has been used as a synonym for electron energy; however, this use is **incorrect** because electron energy is specific to electron ionization, and ionization energy can be applied to any form of ion formation.

Correct: **even-electron ion (EE^+ or EE^-)** – an ion (positive or negative) that contains no unpaired electrons (e.g., CH_3^+ in the ground state). All fragment ions are not necessarily even-electron ions.

Correct: **odd-electron ion ($OE^{+\bullet}$ or $OE^{-\bullet}$)** – an ion (positive or negative) that contains an unpaired electron (e.g., $CH_4^{+\bullet}$). Most single-charge molecular ions (positive or negative) are odd-electron ions. Double-charge molecular ions like $CH_4^{(2+)(2\bullet)}$ are even-electron ions. Because nitric oxide (NO) is a radical, *not* a molecule, the ion produced by the loss of an electron is not a molecular ion (see **radical ion**).

Correct: **fragment ion (X^+, X^-, $X^{+\bullet}$, or $X^{-\bullet}$)** – an electrically charged dissociation product of an ionic fragmentation. A fragment ion can dissociate further, can be positive or negative, and can be an even-electron or odd-electron ion. Fragment ions produced from molecular ions are represented by peak intensities at *m/z* values that correspond to the *m/z* values of fragments of the analyte molecule that are formed by the loss of a neutral molecule or radical or an ion of the opposite charge.

Correct: **imminium ion** – an even-electron ion that is the product of a single-bond cleavage of an ion that contains nitrogen. The imminium ion (usually referred to in EI mass spectrometry) has the form: $RC \equiv N^+ - R'$.

Correct: **immonium ion** – an even-electron ion that is the product of a single-bond cleavage of an odd-electron ion (EI mass spectrometry) or the protonation or cationization of an analyte molecule (ES, APCI, or CI mass spectrometry) that contains nitrogen. The immonium ion has the form: $R_2C = N^+R'_2$.

Correct: **ion** – a particle that results from an atom or molecule that has an unequal number of positive and negative components. Ions result from the removal or addition of electrons; or the addition of negative ionic species such as Cl^-, (etc.); or the addition of protons or other positive ionic species such as Na^+, K^+, Ag^+, (etc.) to neutral atoms or molecules. Ion is derived from the Greek word that means "go" because charged particles go toward or away from a charged electrode. The term *ion* was first used by Michael Faraday in 1834.

Correct: **isobaric** – describes ions or peaks that have the same integer *m/z* value but represent different elemental compositions (e.g., carbon monoxide and ethylene single-charge molecular ions both have a nominal *m/z* value of 28). The terms **isobaric ions** and **isobaric peaks** are correct. Isobaric peaks are peaks with the same integer *m/z* value that represent two different ions. These ions could have different elemental compositions or different isomers (structural or optical) of the same elemental composition (acetone and propanal). Isobar is one of those English words that has two completely unrelated meanings. Isobar is also the term used to describe a line on a weather map that connects two areas of equal atmospheric pressure. This is also called isopiestic.

Correct: **isotope cluster** – a group of peaks close to one another that represent ions with the same elemental composition but a different isotopic composition. In mass spectra of substances containing only C, H, N, O, S, Si, P, and halogens, the lowest *m/z* value peak (not the most intense peak) is the monoisotope peak (e.g., peaks representing the molecular ions $^{12}CH_3^{35}Cl^{+\bullet}$, $^{13}CH_3^{35}Cl^{+\bullet}$, $^{12}CH_3^{37}Cl^{+\bullet}$, and $^{13}CH_3^{37}Cl^{+\bullet}$ constitute an isotope cluster).

Correct: **isotopic ion** – any ion that contains one or more of the less abundant naturally occurring stable isotopes of the elements that make up the structure of the ion (e.g., $^{13}CCH_5^+$).

Correct: **isotopic molecular ion** – a molecular ion that contains one or more of the less abundant naturally occurring stable isotopes of the atoms that make up the molecular structure of the ion.

Correct: **metastable ion** – an ion that dissociates after leaving the ion source and before reaching the detector. An ionization process produces three categories of molecular ions: (1) stable ions – those ions that remain intact for the 100 µs or longer needed to reach the detector; (2) unstable ions – those ions that decompose within $<10^{-7}$ s and are detected as fragment ions; (3) metastable ions – those ions that decompose outside of the ion source within 1–100 µs after their formation. Metastable ions are accelerated from the ion source as one species (precursor ions), undergo decomposition as a consequence of energy deposited during ionization, and are detected as product ions.

Correct: **m/z** – the symbol for the "mass-to-charge ratio" of an ion or peak (*m* = mass of the particle; *z* = number of charges on the particle). Although this term is the official usage as prescribed in the *Current IUPAC Recommendations* and the *ASMS Guidelines*, unfortunately, *m/z* is a mass spectrometry neologism. In SI units, the lowercase letter **m** is the symbol for the meter. The symbol for atomic mass unit is the lowercase letter **u**. Therefore, the SI correct abbreviation for a mass spectral ion or a peak in a mass spectrum should be u/z (u = unified atomic mass unit; z = number of charges on the particle). This term has never been used. The single term m/z is a symbol, *not* an abbreviation. The symbol m/z should never be used with the word "ratio" (i.e., m/z ratio) because "ratio" is part of the definition.

It is important to remember that m/z XX is a property (an adjective) of an ion or a peak in a mass spectrum. m/z XX is not an ion (i.e., [**correct**] "the peak at m/z 91 represents a significant ion"; [**incorrect**] "m/z 91 is a significant ion").

In the case of a molecular ion with a single charge ($M^{+\bullet}$ or $M^{-\bullet}$), the ion results from the loss or gain of an electron from or to the neutral molecule (M). In the case of a fragment ion, the ion could be the result of the formation of a positive or negative ion by breaking a chemical bond. Double-charge ions result in an observed intensity at an m/z value of half the mass of the ion, whereas most m/z values observed in the mass spectrum are equal to the mass of the ion because $z = 1$. ES ions of large molecules are often multiple-charge ions. The difference between two peaks in a mass spectrum should be reported as m/z units (i.e., the difference between the peaks observed at m/z 300 and 271 is 29 m/z units, *not* 29 mass units or 29 u). The symbol **u** refers to the unified atomic mass unit, and the symbol *m/z* refers to the mass-to-charge ratio.

Incorrect: **m/z**, *m / z*, or **M/z** – when used as a symbol for the mass-to-charge ratio. The symbol m/z is always italicized, and its elements are *not* separated by spaces. The m is always lower case, even if m/z starts a sentence. m/z is a symbol, *not* a mathematical formula. The separated presentation and the use of M/z has appeared in print, and the symbol m / z is presented in the latest edition of *The ACS Style Guide* (Dodd 1997). However, this use is not the accepted style in any of the major mass spectrometry journals.

Incorrect: **thomson** – when used as a name for an m/z unit or increment. This neologism was proposed by R. G. Cooks and A. L. Rockwood (Cooks, Rockwood 1991) in honor of Sir Joseph John Thomson (English Nobel laureate in physics, 1906, knighted 1908, 1856–1940) as a term to aid in alleviating the confusion caused by the increased occurrence of multiple-charge ions in mass spectrometry. There has been a reluctance by many to accept the use of thomson because it is applying a "unit" to a dimensionless number, which is what m/z is by definition. Another problem with using Th for mass-to-charge (m/z) is that m/z is an adjective (i.e., a peak was observed at m/z 529). By turning m/z into a unit, the statement "a peak was observed at m/z 529" would be changed to "a peak was observed at 529 Th (thomsons)", which is to say "a peak was observed at 529 mass-to-charge ratios". This latter statement does not make sense grammatically or scientifically.

Still yet another problem with thomson as a substitute for m/z is that the term "Thomson number" has already been assigned by IUPAC for use in fluid dynamics. There are also the physics terms: "Thomson scattering" (named after J. J. Thomson) and "Thomson effect" (a.k.a. Kelvin effect) with a "Thomson coefficient (μ)" named after Sir William Thomson (Lord Kelvin, 1824–1907). However, it appears that the use of thomson as a synonym for an m/z unit has caught on with a fringe faction in the mass spectrometry community. It is listed as "may be used" by *Rapid*

Communications in Mass Spectrometry. The Cooks/Rockwood letter (as well as the reference in *Rapid Commun. Mass Spectrom.*) does err in the use of Thomson rather than thomson, and Dalton rather than dalton (U.S. Government Printing Office 1986). The proposed "Th" abbreviation is consistent with rules for abbreviations.

This term is accepted by *Rapid Commun. Mass Spectrom.* in the Instructions to Authors and is becoming somewhat ubiquitous in this journal since the publication of the first edition of this book.

Incorrect: **m/e** or *m/e* – when used as a symbol for the mass-to-charge ratio when referring to the mass in mass units (u) and the number of charges on the ion. The use of *m/e* would be correct if the mass is reported in kilograms and the charge is in coulombs, which is the case in some mass spectrometry presentations that deal with ion physics. The *m/e* term was the nomenclature used in mass spectrometry prior to 1980 for mass in u and charge number. This usage of m/e is now considered to be archaic. The primary reason for the change is that "e" is the symbol for the charge on a single electron (1.6×10^{-19} columns). This value is not what you want to divide into the mass of the ion! Spectra are seen in older literature with the x axis labeled m/e. In this older literature, *m/e* should be considered synonymous with *m/z*. When J. J. Thomson first determined the mass of an electron, the result was an arbitrary value of *e/m*. The mass of the electron was determined relative to the mass of a hydrogen ion (a proton). Neither the value of *e* nor the value of *m* were known until R. A. Millikan's 1909 classic oil-drop experiment established the value for *e* (Millikan 1909). With the value of *e* being known, it was possible to determine the mass, in kilograms, from Thomson's measured *e/m*.

Correct: **molecular ion** ($M^{+\bullet}$ or $M^{-\bullet}$) – describes ions in a mass spectrometer or peaks on a spectrum that result from the ionization of a molecule through the gain or loss of electrons. A molecular ion has the same nominal mass as the molecule from which it was formed. A spectrum with a molecular ion peak has an observed intensity at the *m/z* value that corresponds to the mass of the analyte molecule divided by the number of charges on the ion. To those who will say, "What about the molecular ion of nitric oxide?": Nitric oxide is *not* a molecule—it is a radical. There are ions, radicals, and molecules, which are all differentiated from one another. Even though an ion produced by the loss or addition of an electron from a radical has the same mass as a radical, it is not a molecular ion because the resultant ion is not from a molecule. There can be multiple molecular ions of different mass produced for an analyte because of the multiple possible combinations of isotopes. All single-charge ions that have an *m/z* value equal to the mass of the molecule from which they were produced are molecular ions. All molecular ions will not have the same *m/z* value as the mass of the molecule from which they were produced because a molecular ion can have multiple charges.

*Note: In mass spectrometry, the symbol used to designate the molecular ion is $M^{+\bullet}$ or $M^{-\bullet}$; however, the second edition of the "Gold Book" (*Compendium of Chemical Terminology*, IUPAC Recommendations) published in 1997, under the definition for "radical ion" (p 336), states, "Unless the position of unpaired spin and charge can be associated with specific atoms, superscript dot and charge designation should be placed in the order •+ or •− suggested by the name "radical ion" (e.g., $C_3H_6^{•+}$)."

The concluding paragraph for the definition of "radical ion" states, "In mass spectroscopic [sic] usage, the symbol for the charge precedes the dot representing the unpaired electron." This is another example of the conflicts that exist within IUPAC nomenclature rules. In a Note attached to the current definition, the IUPAC document states, "In previous versions of this *Compendium*, it was recommended to place the charge designation directly above [below in mass spectrometry] the dot ($M^{•}_+$ or $M^{+}_•$). However, this format is now discouraged because of the difficulty of extending it to ions bearing more than one charge and/or more than one unpaired electron."

Incorrect: **molecular ion** − when used to describe the adduct ion produced from a molecule and an ion (e.g., $[M + H]^+$, MH^+, $[M + Na]^+$, etc.).

Correct: **oxonium ion** − an even-electron ion that is the product of a single-bond cleavage of an odd-electron ion that contains oxygen where the original site of the charge is on the oxygen atom. The oxonium ion has the form: $R_2–C=O^+–R'$.

*Incorrect: **parent ion ($P^{+•}$)** − when used to refer to the molecular ion. The use of this term to refer to the molecular ion was discontinued when the technique of MS/MS was developed. After that time, parent ion was used to describe the ion (either a molecular ion or a fragment ion) formed in a primary ionization (EI, CI, FAB, etc.) that was selected for secondary fragmentation (dissociation) to produce secondary fragments called **daughter ions**. These daughter ions could undergo further-induced dissociation to produce **granddaughter ions**. Because of the gender-specific and anthropomorphic nature of those terms (and the fact that the sex of an ion cannot be determined even when they are turned over), the terms *parent*, *daughter*, *granddaughter*, (etc.) have been replaced with *precursor* and *product*, respectively.

*Incorrect: **parent, daughter, granddaughter, (etc.)** − when used to refer to ions in MS/MS experiments. It has been stated by Gary Glish (Glish 1992) that the terms *parent* and *daughter* used to describe ions in MS/MS probably had its origin in the analogy with the parent/daughter terminology used to describe nuclear disintegration and other similar relations. He further states, "This terminology is quite appropriate based on definitions in Webster's dictionary, achieves the goal of conveying the desired information in succinct and concise manner, and the genealogy relationship should be easy for scientists (and nonscientists) outside the field [of mass spectrometry] to grasp." However, he agrees with an editorial by Maurice M. Bursey (Bursey 1992) which states that the term *daughter ion* is found to be offensive to some mass spectrometrists, and "whoever continues to use a term after learning that it is offensive is rude." Jeanette Adams, Emory University, Atlanta, Georgia (Adams 1992), states that daughter, granddaughter, and great-granddaughter are "archaic gender-specific terms", and that parent ion and progeny fragment ions are "anthropomorphic". She says that because ions are "things" and are incapable of either sexual or asexual reproduction, they can neither be mothers, fathers, daughters, or sons. She further quotes *The ACS Style Guide* (Dodd 1986): "...discourages the use of gender-specific language in ACS publications (pp 103–105)." It is obvious that she and many others find all these terms to be offensive; therefore, they should not be used.

The *Current IUPAC Recommendations* and the *ASMS Guidelines* state that product ion is synonymous with daughter ion, and precursor ion is synonymous with parent ion. The gender-specific terms *daughter ion* and *parent ion* are *not* considered correct by this author. In China, the ions referred to as daughter ions are called *son ions* because of the importance of the male progeny in the Chinese culture.

*Correct: **precursor ion** and **product ion** – terms that discuss the ions in an MS/MS experiment. The *Journal of the American Society for Mass Spectrometry* lists these two words in "Standard Definition of Terms Relating to Mass Spectrometry" (Price 1991) as being synonymous with parent ion and daughter ion, respectively. However, in the same issue, Gary Glish (Glish 1992) suggests the use of parent ion and product ion to describe these two species. Because the terms *parent ion* and *daughter ion* are considered to be offensive by some people, the terms *precursor ion* and *product ion* should be used. Referring to the products of a product ion (granddaughter), the term ***x*-generation product ion** should be used. The *Current IUPAC Recommendations* and the *ASMS Guidelines* state, "This term [precursor ion] is synonymous with parent ion." This synonymous relationship is *not* considered correct by this author.

Correct: **protonated molecule** – an adduct formed by combining a proton (H^+) and a molecule (M) to give MH^+ or $[M + H]^+$. A protonated molecule is the product of a gas-phase ion/molecule reaction or solution chemistry that results in a positive ion that has a single charge and a mass 1 u greater than the molecule. A protonated molecule can have more than a single charge that results from the addition of more than a single proton. A protonated molecule is an even-electron ion.

When representing a protonated molecule, use MH^+, $(M + H)^+$, or $[M + H]^+$. Do not use $[M+H]^+$ (no space between the three bracketed elements). The same convention is true for other ions that are formed by adding inorganic ions to a molecule (i.e., $[M + Na]^+$, $[M + K]^+$, etc.) and for ions produced by a hydride abstraction ($[M - H]^+$) or that are formed by the loss of hydrogen ($[M - H]^-$). Omission of the spaces between the mathematical sign on the letters can lead to confusion. Does M–H mean a molecule less a proton or a molecule with an added proton?

Some ideas have been suggested to deal with **cationization** of molecules (Bursey 1992). Generally, these rules involve the use of the stem word used in English for the added element (i.e., sodiuation as opposed to natriation, or silveration as opposed to argentation). However, even the term *cationization* is somewhat questionable with the recommendation that cation *not* be used to describe positive ions in mass spectrometry.

Incorrect: **protonated molecular ion** – when used to describe a molecule that has been protonated. This term implies that the molecular ion (a positive-charge species) has reacted with a proton (another positive-charge species). For this reaction to happen, the two species would have to come in contact with one another. This event is unlikely because like charges repel one another.

Correct: **positive ion** and **negative ion** – terms that describe the charge state of an ion. A *positive ion* is an atom, radical, molecule, or part of a molecule that has one or more fewer electrons than it has protons. A *negative ion* is an atom, radical, molecule, or part of a molecule that has one or more electrons than it has protons. According to the *Current IUPAC Recommendations*: In mass spectrometry, negative ions should not be referred to as anions; and positive ions should not be referred to as cations because of the connotations of these two terms in solution chemistry. The use of the term *mass ion* is also *not* considered correct in the *Current IUPAC Recommendations*.

Incorrect: **positive-** or **negative-charge** – when used to refer to the charge on an ion. *Do not* use positive- or negative-charge to describe the ion. Refer to ions as *positive ions* or *negative ions*—the sign of the charge. Ions are charged particles.

Correct: **principal ion** – a term usually reserved to describe ions that have been artificially isotopically enriched in one or more positions. These ions may be either molecular or fragment ions. Examples of principal ions would be $CH_3{}^{13}CH_3^{+\bullet}$ (a molecular ion having a mass of 31) and CH_2D^+ (a fragment ion having a mass of 17). The exact definition of a principal ion in the *Current IUPAC Recommendations* is "…a molecular or fragment ion which is made up of the most abundant isotopes of each of its atomic constituents." For the case of species that are not the result of an artificial isotopic enrichment, the principal ion could be another description of a *monoisotopic* or *nominal mass ion*; or in the case of high-mass ions, the *most abundant ion* in an isotope cluster.

Correct: **radical ion** or **odd-electron ion** ($OE^{+\bullet}$) – two terms that can be used interchangeably to describe ions that contain an unpaired electron; therefore, these ions are a radical and an ion. In mass spectrometry literature, the *Current IUPAC Recommendations* states that these ions are represented by placing a superscript dot following the superscript symbol for the charge (i.e., $C_2H_6^{+\bullet}$ and $SF_6^{-\bullet}$). These ions formerly were represented as $X^{+\bullet}$. Placing the dot below the + sign, however, could lead to confusion when there are multiple charges (i.e., $X^{(2+)(2\bullet)}$). The *IUPAC Compendium of Chemical Terminology* (1987) lists an alternate form where the dot precedes the sign. This type of presentation is used in the organic and inorganic literature.

Incorrect: **radical cation** or **radical anion** – when used as a synonym for a *positive ion* or a *negative ion*, respectively. Although these terms have been used for years in mass spectrometry, they are no longer recommended by IUPAC (see **positive ion** and **negative ion**).

Correct: **rings plus double bonds** – an expression used to describe a method of determining the number of rings and/or double bonds in an ion. The rings-plus-double-bonds calculation results are also known as the **degrees of unsaturation** and **hydrogen-deficiency equivalents**. However, in mass spectrometry, the rings-plus-double-bonds terminology is the preferred:

$$R + db = X - 1/2Y + 1/2Z + 1$$

where X is the number of C and Si atoms, Y is the number of H and halogen atoms, and Z is the number of N and P atoms. Atoms of O and S do not enter into the calculation. Double bonds associated with the higher valence states of P and N (valence = 5) or S with a valence of 4 or 6 are not determined with this equation.

*Correct: **single-**, **double-**, **triple-**, or **multiple-charge** – terms used as adjectives to describe the number of charges on an ion. They are expressed as hyphenated words. For example, "The mass spectra of aromatic hydrocarbons can exhibit peaks that represent double-charge ions." This description means the ions have two charges.

*Incorrect: **multiple-charged**, **multiply-charged**, or **multiply charged** or words used to specify the number of charges like **single-charged**, **doubly-charged**, **or triply charged** – when used to describe ions with multiple charges. Ions are charged particles; they are not charged (i.e., a charge is not added to the ion). A charged ion is the same as having a negative deficit. The number and sign of the charge can be changed, but to be an ion means to have an electrical charge. To say an ion is "doubly charged" means an ion was charged once (not likely, because only the sign and the magnitude of the charge on an ion can be changed); then it was charged a second time. These events would result in a particle with three or more or less charges, depending on the sign of the ion's original charge and the sign of the subsequently added charges. The *Current IUPAC Recommendations* uses singly-, doubly-, triply-charged, and so forth. This usage is considered to be grammatically incorrect by this author.

*Incorrect: **quasi-molecular ion** – when used to describe an ion that represents the intact molecule. Maurice M. Bursey said, "Show me a quasimolecule, and then I shall agree you can ionize it" (Bursey 1991). This term was originally used to indicate an ion with the approximate, but not the exact, mass of the molecule. An example would be the $[M + 1]^+$ formed in the protonation of a molecule under chemical ionization conditions. The term was later used to describe any ion formed by a process that did not generally disturb the structure of the molecule. In *positive-ion mass spectrometry*, examples would be $[M + 1]^+$, $[M - 1]^+$, $[M + Na]^+$, (etc.); and in *negative-ion mass spectrometry*, $[M - 1]^-$, $[M + Cl]^-$, (etc.). The use of quasi-molecular ion is, however, recommended by IUPAC.

Incorrect: **pseudo-molecular ion** – when used to describe the same type of ion as described by quasi-molecular ion and has the same degree of ambiguity. Both of these terms fall into the class of words "that mean what the user wants them to mean". Therefore, they can mean different things to different users. Two other terms have been reported to describe these types of ions since the first edition of this book. They are **molecular-related ion** and **near-molecular ion**, both of which are **incorrect**.

*Correct: **resolving power** – the term used to define the ability of a mass spectrometer to separate ions of two different *m/z* values. The resolving power of the mass spectrometer is defined as **M/ΔM** where **M** is the *m/z* value of a single-charge ion, and **ΔM** is the difference between M and the value of the next highest *m/z* value ion that can be distinguished (separated) from M in *m/z* units. Resolving power is determined from the measurement of mass spectral peaks and should be reported with the method by which ΔM was determined (i.e., full width at half-maximum height [FWHM], 10% valley [two adjacent mass spectral peaks of equal height overlap so that the height from the baseline to the top of the overlap valley is 10%] method, 50% valley method, etc.). The FWHM results in a value for resolving power that is twice that obtained with the 10% valley method. It should be noted that using the mass spectral peak width at 5% of its full height is considered to be equivalent to the 10% valley method as illustrated in **Figure 10**.

*Incorrect: **resolution** – when defined in the same way as *resolving power*. Resolution is the inverse of resolving power and expressed as **ΔM** at **M**. Although resolving power is a large number and is associated with a "valley" or mass spectral peak width, resolution can be a small number. Resolution is the measure of a separation of two mass spectral peaks. Resolution is often reported in terms of parts per million (ppm), which can be misleading depending on whether or not it is associated with a specific *m/z* value (Gross 1994). In physics, resolution is used to describe the separation of a vector into its components or the amount of information or detail revealed in an image procedure. In chemistry, resolution is used to describe the separation of a "racemic mixture" into its optically active components or components in a chromatographic separation.

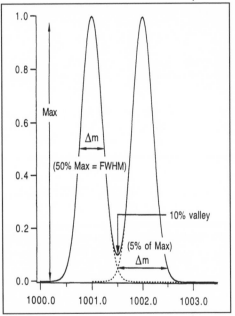

Note: The *Current IUPAC Recommendations* and *ASMS Guidelines* define resolution and resolving power in the opposite way than how they are defined in this document. However, the "MS Terms and Definitions" that appears on the ASMS Web site (http://www.asms.org) under Items of Interest uses language that is in agreement with the above definitions.

Figure 10. Illustration of the peaks used to calculate the resolving power of a mass spectrometer showing the location of the FWHM, the 10% valley, and the 5% valley.

Note: Keith Jennings (University of Warwick, Coventry, United Kingdom) in a private conversation during the 15th International Mass Spectrometry Conference in Barcelona, Spain, regarding the first edition of this book stated, "The term resolving power is a property of the mass spectrometer and resolution is a property of the mass spectrum." Another way of saying the same thing would be that resolving power is a property of the instrument, and resolution is a property of the mass spectral peak.

Incorrect: **satellite ion** or **satellite peak** – when used as synonyms for *isotope* or *isotopic*, respectively. Because a satellite usually "revolves" or "is" around or in some way associated with some primary object, the use of satellite to describe peaks in a mass spectrum or ions related to an ion of a given *m/z* value could refer to [X – 1], [X – 2], (etc.), as well as those that are possibly involved with isotopes [X + 1], [X + 2], (etc.). This ambiguity is the reason why the word *satellite* should not be used when referring to isotopes.

Correct: **tropylium ion** – an even-electron ion that is the product of a single-bond cleavage of an odd-electron ion that contains a benzyl group (C_6H_5–CH_2–R) and the charge is on the ring. The tropylium ion results from a cleavage between the methylene carbon and the atom adjacent to it followed by the formation of a seven-member ring with a phenyl group and the methylene carbon.

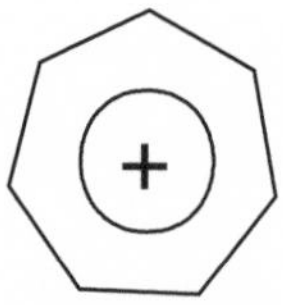

A Tropylium Ion

Correct: **tolyl ion** – an even-electron ion that is the product of a single-bond cleavage of an odd-electron ion that contains a tolyl group (CH_3–C_6H_4–R). The tolyl ion results from a cleavage between the R group and the ring to produce an ion with the form: CH_3–$C_6H_4^+$ where the charge is on the ring.

CH_3

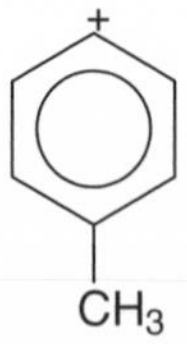

A Tolyl Ion

Note: As examples, the word *isotope* modifies the words *peak*, *cluster*, and *number*. The word *isotopic*, which is an adjective, modifies the word *ion* (author's convention).

MASS

Correct: **atom** – the fundamental particle of a chemical element. Atoms are neutral. They are composed of a nucleus that consists of positively charged particles (*protons*) and neutral particles (*neutrons*), both having about the same mass, and a number of negatively charged particles (*electrons*) that are equal to the number of protons. The electrons exist in energy shells (orbitals) around the nucleus. All but 19 of the known elements have multiple naturally occurring stable *isotopes* (particles that have the same atomic number but different mass numbers). The *nominal mass isotope* of hydrogen is the only atom that does not contain neutrons. Electrons have a mass that is 1/1837 that of a proton. Atoms are characterized by their *atomic number* (number of protons in the nucleus) and *mass number* (an indication of the number of protons and neutrons in the nucleus). The proton was discovered in 1919 by Ernest Rutherford (Rutherford 1920), 1871–1937, New Zealand physicist. The neutron was discovered in 1932 by Sir James Chadwick (Chadwick 1932), 1891–1974, British physicist. The electron was discovered in 1897 by J. J. Thomson (Thomson 1897), 1856–1940, English physicist who calls them "corpuscles". The term *proton* was proposed by Rutherford in honor of William Prout's (1785–1850, British chemist and physiologist) 1815 hypothesis that the elements are composed of multiples of a single "protyle" unit that he identified with hydrogen (the *whole number theory*). Fredrick Soddy (1877–1956, English physical chemist) proposed the term "hydrion" to stress the proton's identity with a hydrogen ion. The term *electron* is attributed to the Irish physicist G. Johnson Stoney (Stoney 1891), 1826–1911, who first used the term in an 1891 address before the Royal Dublin Society. Although Chadwick used the term *neutron* for the nuclear component of an atom that had no charge and a mass equal to that of the proton, the term had been coined two years earlier by Wolfgang Pauli (1900–1958, Austrian-born Swiss physicist) to describe a hypothetical neutral *massless* particle required for the justification of the apparent energy conservation violation in radioactive β decay. Because Pauli had never used this term in print, the tendency was to use Chadwick's heavy neutron definition. Enrico Fermi (1901–1954, Italian-born U.S. physicist) proposed calling Pauli's particle the "neutrino" or "little electron" in Italian. Elements with an even atomic number are generally more abundant than elements with an odd atomic number (Harkin's rule). The word *atom* is derived from the Greek word *átomos*, which means undivided. The principle of the atom and the word was first proposed and used by the Greek philosopher Democritus (470–380 B.C.) and later by John Dalton (1766–1844, English chemist).

Correct: **u** – the symbol for a mass unit or the *unified atomic mass unit*. This symbol represents 1/12th the mass of the most abundant naturally occurring stable isotope of carbon. The term **dalton** (symbol **Da**) is used in biochemistry and is accepted in mass spectrometry; however, it has not been approved by the Conférence Générale des Poids et Mesures. The mass of a given particle is the sum of the atomic mass (in daltons) of all the atoms of the elements composing it. The use of the symbol **u** is referenced in *Quantities, Units and Symbols in Physical Chemistry*, 2nd ed. (Mills 1993).

Incorrect: **amu** – when used as the symbol for mass unit. The *Current IUPAC Recommendations* no longer recommends amu as the symbol for mass. The amu symbol for a unit of mass was used when the standard for mass was based on "oxygen 16". Physicists reported mass based on the most abundant naturally occurring stable isotope of oxygen (^{16}O: established by Francis William Aston [1877–1945] in 1929 after his discovery that oxygen was composed of three different isotopes, two of which had a higher mass [^{17}O and ^{18}O] than the most abundant isotope), which was assigned an exact mass of 16 (1 amu = 1/16 of the mass of ^{16}O). This definition was the basis of the *physical atomic mass scale*. Chemists used amu to define a unit of mass as 1/16 the atomic weight (the average atomic mass) of oxygen (officially established in 1905 on the suggestion of the Belgium chemist, Jean Servais Stas, 1813–1891). This definition was the basis for the *chemical atomic mass scale*. The two scales differed by a factor of 1.000275 (physical > chemical).

An **atomic mass** (a.k.a. **atomic weight** in American and British English) is the weighted average of the mass of the naturally occurring stable isotopes of an element, and oxygen has three such isotopes. The atomic mass of oxygen was an absolute value of 16 by definition. The chemical atomic mass scale made the determination of the atomic mass of newly discovered elements easy by forming their oxides. The atomic mass is used in the calculation of the average mass.

To eliminate the ambiguity between the physical and chemical standards, the standard of a single mass unit as 1/12 the most abundant naturally occurring stable isotope of carbon (^{12}C) was adopted in 1960 by the International Union of Physicists at Ottawa and in 1961 by the International Union of Chemists at Montreal. This standard is based on the independent recommendations of D. A. Olander and A. O. Nier in 1957. The symbol for the *unified atomic mass unit* was established as **u**, *not* **μ**, which appears in the *ASMS Guidelines* and in the 4th edition of *Interpretation of Mass Spectra* (McLafferty 1993).

Prior to the oxygen standards, the basis for atomic mass had been set ca. 1805 by John Dalton (1766–1844) as 1 for the lightest element, hydrogen. In 1815, the Swedish scientist Jöns Jacob Berzelius (1779–1848) set the atomic weight (relative atomic mass) of oxygen to 100 in his table of atomic masses; however, the Berzelius standard of mass was not accepted. The Stas recommendation of setting the "oxygen 16" standard allowed hydrogen to retain a mass close to 1, thereby keeping Dalton's scale somewhat intact.

Science, especially in the United States and the United Kindom, has had to deal with the difference between the use of the terms *mass* and *weight*. These terms are often *incorrectly* used interchangeably. *Mass* is the measure of the quantity of matter, whereas *weight* is the force of a mass due to the acceleration of gravity. The mass of an object will remain the same whether it is on the earth or on the moon; however, its weight will differ by a factor of 6 because the acceleration of gravity differs by a factor of 6 between the earth and the moon. Although extensively used in much of science, the term *atomic weight* is actually **incorrect** because the reference is to a quantity, *not* a force. To be

correct, the term *atomic mass* should be used. However, atomic mass is used to refer to the mass of individual isotopes; therefore, the differentiating term for the weighted average of the naturally occurring stable isotopes is *atomic mass* (a.k.a. *atomic weight*). The use of atomic weight is not very common in mass spectrometry because mass spectrometry deals with the mass of the isotopes, *not* the atomic mass of elements. The lack of differentiation between mass and weight in common language, especially in the southern part of the United States, led to another attempted propagation of some rather bizarre neologism in a 1960s general chemistry text book written by two individuals at the University of Tennessee, Knoxville, Tennessee. The authors of this book gave a somewhat egalitarian treatment to the terms **atomic mass unit**, **atomic weight unit**, and **avogram**. Throughout the book, awu was used rather than amu. Their rationale was: "There is no general agreement as to what this (atomic mass) unit should be called. Although it is also called atomic mass unit and avogram, we shall use the term atomic weight unit because of its descriptive nature" (Keenan 1961).

Another important consideration with regard to the atomic mass of carbon is the difference in the abundance of the carbon-13 isotope depending on the source of the carbon, petroleum or biological. This **carbon average mass variation** will have to be considered in the calculation of high-mass substances being measured by mass spectrometry. Carbon associated with biological sources of organic compounds has a higher ratio of ^{13}C than does the carbon in organic compounds whose origin is petroleum. Because mass spectrometry is now involved with the detection of ions with mass high enough that the various isotope peaks cannot be resolved due to instrument limitations, the average mass (atomic mass) of the ion is measured. The problem of these high-mass ions is further aggravated by the fact that the large number of carbon atoms in these ions makes it impossible to detect an ion with no, one, or two atoms of ^{13}C. Therefore, only the average mass can be measured. The calculated average mass for an ion representing an intact molecule (molecular mass) of biological origin can vary from the accurately measured mass by a significant amount. A good example for this is the difference in the calculated and measured average molecular mass of myoglobin. The calculated value is ~16951.3 Da, and the measured value is ~16951.5 Da.

Inappropriate: **A.M.U.** or **a.m.u.** – when used as abbreviations for atomic mass unit. However, they should be avoided because of the confusion with the symbol **amu** (see **amu**, above). The symbol for the unified atomic mass unit is **u**.

Note: The use of **amu**, **a.m.u.**, **A.M.U.**, or any other term used for mass (**dalton**, **Da**, **u**, or **mass**) as a label for the abscissa of a mass spectrum is **incorrect** when the abscissa represents values of mass-to-charge ratios of ions. Only use terms that represent units of mass when the *m/z* values have been converted to mass; never use amu, a.m.u., or A.M.U. to represent mass.

In the older literature, the abscissa of the mass spectrum was often labeled "mass". This type of presentation was used because, in nearly every case, the number of charges on the ion was one. After the development of the electrospray interface and the analysis of bio-macromolecules with the presence of multiple-charge ions became widespread, the use of "mass" on the abscissa of the mass spectrum (although always ill-advised) became more problematic and potentially confusing.

Care must also be taken in the use of the terms **molecular weight** and **molecular mass**. The *molecular weight* (a.k.a. relative molecular mass) of a molecule is based on the *atomic mass* (relative atomic mass to several decimal places) of its atoms, whereas the *molecular mass* is based on either the *nominal, monoisotopic,* or *isotopic masses* of its elements.

Correct: **atomic number** (a.k.a. **proton number**) – the number of protons in an atom of a element. All elements have a unique atomic number that determines the position of the element in the periodic table. The IUPAC symbol for the atomic number is **Z**.

Correct: **average mass** – the calculated mass of a molecule, ion, or radical based on the atomic mass (relative atomic mass) of the elements from which it is composed. An *atomic mass* is the weighted average of the naturally occurring stable isotopes of an element. Atomic mass of carbon:

$$(12.000 \times 0.989) + (13.0034 \times 0.011) = 12.0110 \text{ (see } \textbf{Figure 11})$$

Incorrect: **atomic weight** – when used to describe the mass of individual isotopes of an element.   The integer mass of the most abundant naturally occurring stable isotope of an element is the *nominal mass*. The exact mass of the most abundant naturally occurring stable isotope of an element is the *monoisotopic mass*. Some authors will refer to the atomic weight (or atomic mass) of an element (weighted average of the naturally occurring stable isotopes) while also referring to the atomic weight (or atomic mass) of a specific isotope of an element. This is incorrect.

Correct: **calculated exact mass** – the mass determined by summing the mass of the individual isotopes that compose a single ion, radical, or molecule based on a single mass unit being equal to 1/12 the mass of the most abundant naturally occurring stable isotope of carbon. The exact mass values used for the isotopes of each element are recorded in tables of isotopes. The mass (atomic mass) of an element that appears in the periodic table is a weighted average of these exact mass values of the naturally occurring stable isotopes of that element. If the mass is calculated with the exact mass value of the most abundant naturally occurring stable isotope of each element in the ion, radical, or molecule, then the *calculated exact mass* is the same as the *monoisotopic mass*.

Incorrect: **exact mass** – when used to mean the experimentally determined (measured with a mass spectrometer) mass of an ion, radical, or molecule. Exact mass has been used to describe measured and calculated mass, leading to a degree of confusion. Exact mass should never be used to mean an accurately measured mass.

Correct: **measured accurate mass** – an experimentally determined mass of an ion, radical, or molecule that allows the elemental composition to be deduced. For ions with a mass ≤200, a measurement to ±5 ppm is sufficient for the determination of the elemental composition. The term *measured accurate mass* is used when reporting the mass to some number of decimal places (usually a minimum of 3). A measured mass should be reported with a precision of the measurement (Gross 1994).

Incorrect: **accurate mass** – when used to mean the calculated mass of an ion, radical, or molecule. Accurate mass has been used to describe measured and calculated mass, leading to a degree of confusion. Accurate mass should never be used to mean calculated mass. The use of accurate mass in the reporting of a measured mass should be used with the word *measured*—a *measured accurate mass.*

Correct: **isotope** – an atom of the same element that differs in *mass number* (the sum of the protons and neutrons that compose the nucleus of the atom) due to the difference in the number of neutrons in the nucleus (see **nuclide**). Elements with even atomic numbers have a tendency to have more naturally occurring stable isotopes than those with odd atomic numbers. With the exception of potassium, the elements with odd atomic numbers do not have more than two naturally occurring stable isotopes. The term was proposed by Frederick Soddy (1877–1956), the English physical chemist, in 1917 for different radioactive forms of the same chemical species because they are classified in the same place in the periodic table (Gr, *isos*, equal, same + Gr, *topos*, place; that is, "having the same place" in the periodic table). Later, Francis William Aston (1877–1945), an English physicist, applied the term to atoms of the same element that had different mass but that did not undergo radioactive decay—stable isotopes. Aston is best known for the discovery of most of the naturally occurring stable isotopes of the first 92 elements in the periodic table. Aston and Soddy were Nobel Prize recipients, and both worked in J. J. Thomson's Cavendish Laboratory.

Correct: **isotope number** – the difference between the number of neutrons and the number of protons in a nuclide. Different isotopes of the same element have different isotope numbers.

Correct: **isotope enrichment** – a process of preparing an analog of a compound that contains one or more atoms of an isotope of a particular element beyond the naturally occurring abundance. These analogs are isotopically labeled compounds. The isotope labeling can be accomplished with either stable or radioactive isotopes.

Correct: **isotope dilution** – the technique of adding a known amount of an isotopically labeled variant of an analyte as an internal standard. This technique will lower the detection limit for the analyte if the isotopically labeled substance is added prior to sample processing. Care must be taken when using this technique to compensate for the amount of nominal mass material and any contribution to the internal standard peak intensity that may be due to the naturally occurring stable isotopes of the analyte.

Correct: **isotopic mass** – the mass of any isotope.

Correct: **isotopomer** – because there should be one and only one definition for a term (in order to prevent confusion), isotopomers are chemical compounds differing from each other only in respect to the quantity of an isotope of one of the constituent atoms (e.g., $^{12}C^{16}O$, $^{13}C^{16}O$, $^{13}C^{17}O$, $^{12}C^{18}O$, $^{13}C^{18}O$ are isotopomers of carbon monoxide). All naturally occurring organic molecules contain mixtures of isotopomers with ^{12}C, ^{1}H, ^{16}O, ^{32}S, ^{14}N, (etc.) representing the dominant naturally abundant isotopomer (Markey 2000). Another definition of an isotopomer that fits within this general definition: an **isotopically enriched analog** of a compound (molecule, ion, or radical) is an isotopomer of that compound (i.e., CD_3Cl is an isotopomer of CH_3Cl).

Incorrect: **isotopomer** – when used to describe: (1) molecules of a particular compound (ions or radicals) that have been *isotopically enriched* and that have the same number of isotopes, but the isotopes are located in different positions (Freiser 1996) (i.e., $DCH_2–CH(OH)–CH_3$ is an isotopomer of $CH_3–CH(OD)–CH_3$); (2) molecules of the same compound (ions or radicals) that have the same mass but have different isotope enrichments (e.g., 1 mass greater than the native molecule [ion or radical] due to enrichment with ^{2}H, ^{13}C, and ^{15}N would represent three isotopomers of the same compound); (3) isotopically labeled variants of different mass (e.g., molecules of the same compound with different numbers of isotopically enriched atoms of the same element).

Incorrect: **isotopologies** – when used to describe molecules (ions or radicals) of a particular compound (having different numbers of isotopically enriched atoms of a particular element) (Freiser 1996). They would be *isotopomers*.

Correct: **mass defect** – in mass spectrometry, the difference between the exact mass of an atom, molecule, ion, or radical and its integer mass. In physics, the mass defect is the amount by which the mass of an atomic nucleus is different than the sum of the masses of its constituent particles.

Correct: **mass number** – the sum of the total number of protons and neutrons in an atom, molecule, ion, or radical. It is the **nucleon number** with the symbol **m**. This number is an integer and can be used interchangeably with m/z values in unit-resolution mass spectra where the charge number of the ion is 1. However, this practice is not recommended because of the possible confusion that may result. The mass number of isotopes of two different elements can be the same. Although the two elements would each have a unique number of protons, the number of neutrons in each element could be such that both elements had the same mass number. Zirconium has an *atomic number* of 40, and molybdenum has an *atomic number* of 42; however, both have an isotope with a *mass number* of 92 (^{92}Zr and ^{92}Mo). The IUPAC symbol for mass number is **A**.

Correct: **molecule** – a group of atoms that are chemically bonded to one another. Molecules are neutral due to an equal number of positive and negative charges and are characterized by an even number of electrons. The bonds can be covalent (the sharing of a pair of electrons) or ionic (attraction of opposite charges). Particles that have an odd number of electrons that do not have a charge are radicals (nitric oxide [NO] is a radical, *not* a molecule).

A molecule is composed of a unique set of atoms and has a specific geometry. Molecules can be joined together either chemically or physically to form larger molecules or clusters that consist of discrete units of the original molecule. In the general sense, these larger molecules resulting from the combination of a group of the same smaller molecules are called **polymers**. The smaller molecules that are used to form the polymers are **monomers** (*mono-* from the Greek *mons-*, single, alone; *-mer* from the Greek *meros*, part, the final results of many divisions of the whole). A low molecular mass **homopolymer** resulting from an additional polymerization of a few molecules of the same monomer is a **telomer** (e.g., telomers are additional **dimers**, **trimers**, and **tetramers** of C_2F_4). Clusters of molecules where two or more molecules exist as a discrete entity are also called dimers, trimers, tetramers, and so forth. Condensation polymers of a few monomers are called **oligomers**; *olig-* from the Greek *oligos* meaning a few, little (see **oligomer** in the **Biochemical Terms in Mass Spectrometry** section). Polymers can be stereospecific with **isotactic** or **syndiotactic** configuration. An **atactic** polymer has no regular geometric arrangement along the chain.

Defining a molecule as a "group of atoms" prevents the possibility of monoatomic molecules such as the case with the noble gases. Another, and perhaps better because it is more inclusive, definition of a molecule is "*a molecule, the smallest particle of matter that can exist in a free state*". This definition does allow for the existence of the monoatomic molecules of elements such as He, Ar, Xe, (etc.). Oxygen atoms cannot exist in a free state without external forces; therefore, the smallest particle that can exist in a free state that contains oxygen must also contain other atoms (i.e., H_2O, O_2, etc.).

Correct: **most abundant mass** – represented by the most intense peak in an isotope cluster for an ion (e.g., the [X + 2] peak for an isotope cluster that represents an ion that contains only carbon, hydrogen, and four atoms of chlorine).

Correct: **monoisotopic mass** – the exact mass of the most abundant naturally occurring stable isotope of an element. The monoisotopic mass of an ion, radical, or molecule is the sum of the monoisotopic mass of the elements in its formula (e.g., $C_3H_6O^{+\cdot}$ has a monoisotopic mass of 58.0417). The monoisotopic mass of an element is not necessarily the lowest mass naturally occurring stable isotope. For the common elements in organic mass spectrometry (C, H, O, N, S, Si, P, K, Na, and the halogens), the lowest mass isotope is the monoisotopic mass isotope (see **Figure 11**).

Correct: **molecular mass** – describes the mass of a molecule or molecular ion. The molecular mass can be expressed in terms of nominal mass, monoisotopic mass, or atomic mass. When expressed in terms of the atomic mass of the elements of the molecule, the term should be **relative molecular mass** (M_r). Molecular mass should not be used to describe the mass of a fragment ion, adduct ion, or radical because they are not molecules.

Correct: **molecular weight** – describes the mass of a molecule based on the atomic mass (weighted average of the naturally occurring stable isotope) of the atoms that composed it. This number of expressed grams is the **gram molecular weight** of a mole (6.02×10^{23} molecules) of the pure molecule. The molecular weight (a.k.a. *molecular mass*) is the *relative molecular mass*. Ions and radicals do not have molecular weight because they are not molecules.

Correct: **neutral loss** – describes a radical or molecule that is lost from an ion to produce an ion of lower mass. Neutral losses should be reported as units of mass because they represent species that have no charge. Neutral losses are also referred to as **dark matter**.

Correct: **nitrogen rule** – Any common organic molecule or odd-electron ion ($OE^{+\cdot}$) [neither an even-electron ion (EE^+) nor radical] containing C, H, O, S, Si, P, and/or halogens that has an odd number of nitrogen atoms has an odd nominal mass.

Correct: **nuclide** – an atom characterized by its atomic number, mass number, and nuclear energy state. *Nuclide* has incorrectly been considered as a synonym for *isotope*. Isotope refers to different nuclides of the same element. Two nuclides can have the same mass number but be different elements.

IUPAC Statement: In Bulletin #11 (dated August 28, 2000) from the ATOMIC MASS DATA CENTER, an admonishment was issued regarding the "…misuse of the word 'isotope' instead of 'nuclide'…." The following paraphrased definition was included:

Isotopes, isotones, isobars refer to a series of nuclides having the same value for Z (*atomic number*), N (*neutron number*), A (*mass number*), respectively. Therefore, it is correct to speak about isotopes of sodium, or isotones at N = 82, or isobars at mass 144. Nuclides designate one specific combination of neutrons and protons (e.g., ^{87}Rb).

Bulletin #11 further states that the SUNAMCO commission of IUPAC pointed out the misuse in a 1987 document (*Symbols, Units, Nomenclature and Fundamental Constants in Physics*; p 15).   2.5.1 Nuclides of this document reads:

"A species of atoms identical as regards to atomic number (proton number) and mass number (nucleon number) should be indicated by the word 'nuclide', *not* by the word 'isotope'. Different nuclides having the same mass number are called 'isobaric nuclides' or 'isobars'. Different nuclides having the same atomic number are called 'isotopic nuclides' or 'isotopes'...."

In the opinion of this author, such a pedantic view of whether ^{12}C is a nuclide of carbon rather than an isotope of carbon places too great of a burden on a presenter. ^{12}C has been referred to as an isotope of carbon by mass spectrometry from mass spectrometry's earliest beginnings, and the use of *nuclide* and *isotope* as synonyms is permissible and should continue.   Therefore, this is another case of conflict with IUPAC regulations.

Correct: **nominal mass** – the integer mass of the most abundant naturally occurring stable isotope of an element. The nominal mass of an ion, radical, or molecule is the sum of the nominal masses of the elements in its elemental composition (e.g., the nominal mass of $C_3H_6O^{+\cdot}$ is 58). The nominal mass of an element is equal to the *mass number* of the most abundant stable isotope of an element. The lowest mass isotope is not necessarily the nominal mass isotope.

The lowest mass isotope of Hg is 196 (0.1%). The nominal mass is 202 (29.52%). Hg has seven naturally occurring stable isotopes: 196, 198, 199 (16.8%), 200 (23.1%), 201, 202, and 204. The lowest mass isotopes for the common elements in organic MS (C, H, O, N, S, Si, P, K, Na, and the halogens) are the nominal mass isotopes (**Figure 11**).

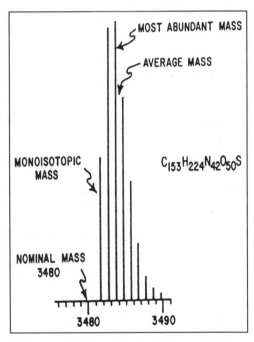

Figure 11.   Illustration of the relationship among the nominal mass, monoisotopic mass, most abundant mass, and average mass of a large-mass single-charge ion. (Yergey 1983)

Incorrect: **nominal mass** – when defined as the integer mass of any isotope of an element or as the integer mass of the lowest mass isotope of an element. Both of these definitions have appeared in mass spectrometry literature. It is important to remember that the integer value for the monoisotopic mass of an ion is *not* the nominal mass of that ion [e.g., the monoisotopic mass of $C_{153}H_{224}N_{42}O_{50}S$ is 3481.59985; its nominal mass is 3480, *not* 3482; the nominal mass is the sum of 153(12), 224(1), 42(14), 50(14), and 1(32)].

Correct: **radical ($X^{\bullet}$)** – a neutral particle that has an unpaired electron (an odd number of electrons).

Correct: **relative atomic mass** – this term has the same meaning as *atomic weight* (i.e., the weighted average of the naturally occurring stable isotopes of an element). The average molecular mass (M_r) of a molecule is based on the sum of the relative atomic mass of the elements that comprise the molecule.

Incorrect: **relative** or **average atomic weight** – when used to describe the weighted average mass of the naturally occurring stable isotopes of an element.

Correct: **whole number rule** – the mass of all atoms (the different naturally occurring stable isotopes of all elements) can be expressed to a high degree of accuracy (in most cases to about 1 part in a thousand). The error expressed in fractions of a mass unit increases with increasing mass. The one exception is hydrogen with an exact mass of 1.007825. There are no atoms of Cl with a mass of 35.6 (the atomic mass of Cl), only those that have a mass of 34.9689 (^{35}Cl) and 36.9659 (^{37}Cl). The difference in the whole number and the actual mass is the *mass defect*. The *whole number rule* was put forth by British chemist William Prout in 1815 and vindicated by F. W. Aston in 1920 (Aston 1924).

Correct: **compound** – the term used to describe a group of atoms of two or more elements where the proportions of each element is fixed. Compounds consist of a collection of the same molecules. There are **ionic compounds** such as sodium chloride (Na^+ ions combined with Cl^- ions) and **covalent compounds** such as methanol (H_3COH) where the atoms share electrons with one another in such a way that the outer shell of all the atoms is **valence** satisfied (has eight electrons in its outermost energy shell). The components of a compound cannot be separated by physical means, and the properties of a compound are different from those of the elements from which it is composed.

What Mass?

What **mass** is measured in mass spectrometry? It is true that mass spectrometry does not measure mass but measures the mass-to-charge ratio of ions. The mass of the ion can be determined by dividing the *m/z* value of the ion by the number of charges on the ion. This is partly why we call the technique "mass spectrometry".

In mass spectrometry, it has often been said that we deal with the mass of the isotopes of the elements, not their atomic masses (a.k.a. relative atomic masses). The atomic mass of an element is the weighted average of the masses of the element's naturally occurring stable isotopes.

In a mass spectrum of single-charge ions representing the molecular ion of bromobenzene obtained using any commercially manufactured electron ionization (EI) GC/MS system, there are clearly four peaks that are about 1 *m/z* unit apart. The peak at the lowest *m/z* value is the **monoisotopic peak** because all the atoms of all of the elements in the ion represented by that peak are only the most abundant naturally occurring stable isotopes of those elements. In this example, the peak at the next highest *m/z* value represents ions of two different possible isotopic compositions: one containing a single atom of ^{13}C and the other containing a single atom of ^2H. The peak 2 *m/z* units higher than the monoisotopic peak represents ions of as many as four different isotopic compositions: one that contains an atom of ^{81}Br, one that contains two atoms of ^{13}C, one that contains two atoms of ^2H, and one that contains one atom of ^{13}C and one atom of ^2H. The peak 3 *m/z* units higher than the monoisotopic peak represents ions of six different isotopic compositions: one atom of ^{81}Br and one atom of ^{13}C, one atom of ^{81}Br and one atom of ^2H, three atoms of ^{13}C, three atoms of ^2H, two atoms of ^{13}C and one atom of ^2H, and two atoms of ^2H and one atom of ^{13}C.

The ions representing the isotopic compositions of $^{12}C_6{}^1H_5{}^{79}Br$ and $^{13}C^{12}C_5{}^1H_5{}^{79}Br$ are separated because the difference in their *m/z* values is ~1 and the mass spectrometer has sufficient resolving power to separate these two ions at this particular *m/z* value. The ions representing the isotopic compositions of $^{13}C^{12}C_5{}^1H_5{}^{79}Br$ and $^{12}C_6{}^2H^1H_4{}^{79}Br$ are not separated because the differences in their *m/z* value is ~0.0107 *m/z* units and the instrument's resolving power is not sufficient to separate these two ions. However, the resolving power does separate the major isotopic compositions of $^{12}C_6{}^1H_5{}^{79}Br$, $^{13}C^{12}C_5{}^1H_5{}^{79}Br$, $^{12}C_6{}^1H_5{}^{81}Br$, and $^{13}C^{12}C_5{}^1H_5{}^{81}Br$. The integer molecular mass (*M*) for this compound is reported as 156 Da. In mass spectrometry, the integer average molecular mass (*M*$_r$) obtained from the atomic masses of the elements constituting the compound's elemental composition of 157 Da would never be used.

Ions have a **nominal mass**. This is the mass of the ion *calculated* based on the integer mass of the most abundant naturally occurring stable isotope of the atoms of the elements composing the ion. In instruments that have unit resolution throughout the *m/z* scale (an ability to separate ions that differ by ~1 *m/z* unit regardless of the *m/z* value), the nominal mass and the measured mass are nearly the same, within the rules for rounding to integer values, for all ions that have a mass of less than about 500 Da. Once the mass of an organic ion (an ion that contains atoms of C and H) gets to a mass greater than 500 Da, **mass defect** (the difference in the actual mass of a nuclide and its integer mass) must be considered. Single-charge ions below 500 Da will exhibit a monoisotopic mass peak at an *m/z* value that is close to that of the ion's nominal mass. Mass defect becomes a significant factor for ions greater than 500 Da primarily because of the mass defect of hydrogen, which is ~0.8% of its integer mass (the actual mass of 1.007825 Da is 0.8% higher than the nominal mass of 1 Da). This value is greater than the mass defect of any other element usually encountered in organic compounds. What makes the mass defect of hydrogen more egregious is the fact that there are about two hydrogen atoms present for every carbon in many organic compounds.

The ion with an elemental composition of $C_{153}H_{224}N_{42}O_{50}S$ has a nominal mass of 3480 Da; however, this elemental composition has a monoisotopic mass of 3481.6 Da and, as a single-charge ion, would be represented by a peak at m/z 3482 in a mass spectrum obtained on an instrument with unit resolution throughout the m/z scale. There are commercially available transmission quadrupole mass spectrometers that do have unit resolution throughout their m/z scale and that will detect ions up to an m/z value of 4000. There are reflectron time-of-flight mass spectrometers (capable of measuring very large m/z values) that have resolving powers (a measured m/z value divided by the difference in m/z values that can be separated; i.e., an ion with m/z 20,000 separated from an ion with m/z 20,001 represents a resolving power [R] of 20,000) of 10,000, 15,000, and 20,000. This means that it is possible to separate the major isotopic compositions from one another for ions of very high mass.

There is another factor that must be considered with respect to the mass spectrum of ions with large mass; and that is the fact that the mass spectrum may not exhibit a monoisotopic peak. Because a large mass ion will contain a large number of atoms of the various elements, there is a higher probability that ions containing one or more heavier isotopes are present than there is that ions containing only the most abundant naturally occurring stable isotopes of the elements are present. In the $C_{153}H_{224}N_{42}O_{50}S$ example, the monoisotopic peak is only about 50% of the intensity of the base peak, which is the X+2 peak in the isotope cluster. Even when there is sufficient resolving power to separate the major isotopic compositions, it may not be possible to determine the monoisotopic mass because the abundance of the monoisotopic ion may be below the level of detection.

It is possible to measure the mass of an ion where the major isotopic compositions cannot be separated because of limits on the resolving power of the instrument. A good example is the mass spectrum of the single-charge protonated molecule of insulin obtained using a MALDI TOF instrument that has a resolving power (R) of ~1000. This ion has an m/z value of ~6K. At m/z 6000, an instrument with R = 1000 will separate ions that have a difference 6 m/z units. This is clearly not sufficient to separate ions representing the major isotopic compositions. So the question is, "What does the m/z value reported for this peak represent?" It represents a weighted average of the individual masses of the major isotopic compositions of the ions of a specific elemental composition. This average is weighted based on the abundances of the ions representing the major isotopic compositions of the elemental composition. Those abundances are a function of natural abundances of the isotopes of each of the elements. Therefore, the value represented by the mass spectral peak is the relative mass of the ion, which can be used to obtain the M_r of the molecule.

When the isotope peaks are not resolved, the m/z value of the ion represents the relative mass of the ion. This is why the "mass" determined for an intact protein may differ from the "mass" calculated from the monoisotopic masses of the amino acids that comprise the protein. Values observed above m/z 500 don't always represent the nominal mass of an ion; m/z values that exceed the resolving power of the measurement represent relative mass instead of monoisotopic mass.

TERMS ASSOCIATED WITH BIOCHEMICAL MASS SPECTROMETRY

Mass spectrometry is increasingly being used in biochemistry and areas related to biochemistry. This field has some terms that are finding their way into modern mass spectrometry jargon. Some of these terms are included in this section. Some of the included terms are not directly related to mass spectrometry; however, they are terms that the mass spectrometrist may encounter and be naive as to their meaning.

Correct: **ADME** – the absorption, distribution, metabolism, and excretion, usually of a drug. Other acronyms that appear in the literature redrawing similar topics are ADMET: ADME and toxicity a.k.a. ADME-tox; and eADMET: early ADMET.

Correct: **amino acid** – an organic compound that contains both an amino and an acid functionality. In biochemistry, there are 20 α-amino acids found in living organisms. A list of these acids can be found in the Tables that make up the material at the end of this book. An α-amino acid has both an amine and a carboxylic acid attached to the same carbon that is on the end of the molecule:

L-Amino Acid   D-Amino Acid

This α-carbon is a chiral center. Therefore, there are two stereoisomers of the amino acids with the exception of the simplest, which is glycine ($H_2C(NH_2)COOH$). Only the L-amino acids are biologically active. Through a condensation of the amino group of one amino acid with the carboxylic acid group of another amino acid, a peptide bond is formed. Polymers produced through a series of these peptide bonds are proteins.

Correct: **C-terminus** (a.k.a. **carboxyl terminus** or **C-terminal group**) – the residue at the end of a peptide that has a free carboxyl group or does not acylate another amino acid residue. The C-terminus may acylate ammonia to give $-NH–CHR–CO–NH_2$ (see **N-terminus**).

Correct: **genome** and **genomics** – the total genetic content of the haploid chromosome set which, together with the pertinent protoplasm, specifies the material foundation of a species. The term is attributed to Hans Winkler (Winkler 1920). The term *genomics* was introduced ca. 1977 by Victor McKusik and Frank Ruddle as the name of a journal. It has a narrow connotation of linear gene mapping and DNA sequencing. It is often used with modifiers such as *functional* and *structural* genomics.

Correct: **high-throughput screening** (**HTS**) – a process for rapid assessment of the activity of samples from a combinatorial library or other compound collection, often by running parallel assays in plates of 96 or more wells; and **ultra-HTS** (**UHTS**).

Correct: **in silico** – the term coined by some clever person to describe the computer-based simulations or mathematical models, relating to biochemistry, behavior, and environment, for a trial carried out, as it were, on a computer chip. This term is most closely related to two older Latin phrases that are ubiquitous in the jargon of biology and biochemistry: *in vivo* and *in vitro*, both of which came into use at the end of the 19th century. The first translates as "in that which is alive" and refers to an experiment carried out within a living organism, such as a drug test on an animal. The second term means "in glass" and is used for experiments that take place in an artificial environment outside the body, such as a test tube or culture dish. The mass spectrometrist will use the term **in silico digestion** to describe the process of having a computer perform a digestion of a protein based on the known behavior of a proteolytic enzyme. An "in silico digestion of a protein" in a protein database using trypsin will generate information on peptide masses obtained that can be compared with mass spectral data in order to identify an unknown protein.

Correct: **isoelectric point (IEP)** (symbol: **pI**) – the point at which a substance, such as a colloid or protein, has zero net electric charge. Usually, such substances are positively or negatively charged, depending on whether hydrogen ions or hydroxyl ions are predominantly absorbed. At the IEP, the net charge on the substance is zero, as positive and negative ions are absorbed equally. The substance has its minimum conductivity at its isoelectric point and therefore coagulates best at this point. In the case of hydrophilic substances in which the surrounding water prevents coagulation, the IEP is at the minimum of stability. The IEP is characterized by the value of the pH at that point. Above the isoelectric pH level, the substance acts as a base; below this level, it acts as an acid. For example, at the IEP, the pH of gelatin is 4.7. Proteins precipitate most readily at their isoelectric points; this property can be utilized to separate mixtures of proteins or amino acids.

Correct: **isotope-coded affinity tag (ICAT™)** – an adduct that covalently bonds to a specific amino acid (cysteine) found in a protein to allow isolation from complex mixtures. The tag is divided into three elements: (1) a **biotin** tag that will allow for the affinity capture of the tagged protein when it is submitted to chromatography using an **avidin** column (68-kDa glycoprotein composed of four identical subunits found in egg whites that will bind with a biotin moiety for each subunit); (2) a linker chain that can be isotopically labeled using either deuterium or ^{13}C; (3) the reactive group, which will bind to and modify cysteine residues in the protein (iodoacetamide alkylation). The technique of isotope-coded affinity tagging was developed by Ruedi Aebersold (Gygi 1999). The technique uses both the stable isotope-labeled and the nonlabeled forms (referred to as light and heavy reagents) to produce pairs of mass spectral peaks (separated by 8 *m/z* units for single-charge ions) for otherwise identical peptides that incorporate the tagging reagent. The technique allows for simultaneous identification and quantitation of a protein and is used to evaluate differential expression of proteins from tissues undergrown under dissimilar biological conditions (e.g., diseased vs healthy tissue or varying nutritional environments). The term **ICAT** is trademarked by the University of Washington in the United States and certain other countries, exclusively licensed to Applied Biosystems Group of Applera Corporation, and should not be applied generically.

In recent years, other similar approaches have been developed that involve labeling at different amino acids, use less-massive reagent tags, and allow for postlabeling cleavage of a portion of the tag. These new strategies differ in the chemistry employed but still rely on the same general principles as ICAT.

Biotin Tag   Linker Chain
(heavy D or light H)   Reactive
Group

Correct: **ladder sequencing** – an order of constituents in a polymer, especially the order of nucleotides in a nucleic acid or of the amino acids in a protein.

Correct: **mass mapping** – a technique, when linked with a database search, that is used to generate a short list of possible identities for an unknown peptide. Mass mapping is based on the fact that proteolytic enzymes will create predictable fragmentation patterns in proteins. These resulting fragments can then be analyzed by mass spectrometry to determine their individual mass. A protein whose sequence is closely related to a known sequence may also be analyzed by this technique.

Correct: **metabolome** and **metabolomics** – the news of the *omics* to invade the language. In a *Commentary* by Joshua Lederberg and Alexa T. McCray in the Web edition of *The Scientist* **2001**, *15*(7), 8, April 2, a list of over 40 scientific words ending in *ome* was included. Metabolome was not one of these words; therefore, it is probably safe to assume that it was coined some time after that date. Metabolomics (not to be confused with **metabonomics**) is the global study of metabolites. Metabonomics specifically refers to the study of changes in metabolic profile of higher organisms in response to stress (which also was not listed in the Lederberg *Commentary*). According to Chris Becker, director of chemistry at SurroMed, Mountainview, California, "Biology encompasses genes, proteins, and metabolites. Any biologist without an agenda to promote will tell you that each of these are important." Therefore, because there is a genome and a proteome, there should be a metabolome.

There have been some recent complaints about the ugliness of the word *metabolomics*. In the above-cited Lederburg *Commentary*, it is stated that Roland W. Brown's *Composition of Scientific Words*, 1954, says, "Words, when they make their debut in scientific or literary society ... should be simple, euphonious, and mnemonically attractive. With the already existence of metabonomics, metabolomics is not mnemonically attractive.   The term *metabolomics* is definitely neither pure nor euphonious."

Correct: **nucleic acids** – polymers of nucleotides where the phosphate group bridges the 3′ position (point of original attachment of the phosphate to the pentose) of the sugar moiety of one nucleotide and the 5′ position of the sugar moiety of another nucleotide. Nucleic acids are composed of three bases that are derived from pyrimidine (six-membered unsaturated ring with N at the 1 and 3 positions), cytosine, uracil, and thymine; and two bases derived from purine (a fused six-membered and five-membered unsaturated ring with N at the 1, 3, 7, and 9 positions), adenine, and guanine (see following structures). The nucleic acid **deoxyribonucleic acid** (**DNA**) is composed of nucleotides containing the purine bases adenine (A) and guanine (G) and the pyrimidine bases thymine (T) and cytosine (C) and 2′-deoxy-D-ribose (the OH on the number 2′ carbon of the pentose residue has been replaced with a H). The same bases are also found in **ribonucleic acid** (**RNA**) with the exception that thymine is replaced by the pyrimidine base uracil (U). In RNA, the pentose moiety has an OH on the 2′ carbon. RNA is the primary constituent of many viruses and is responsible for the self-replication of viruses. There is also a messenger RNA (mRNA) that transmits the coded information contained by chromosomes of the nucleus of a cell to the protein-producing ribosomes of the cytoplasm. Transfer RNA (tRNA) transfers activated amino acids onto the mRNA. In double-stranded DNA, the amount of A is equal to the amount of T, and the amount of G is equal to the amount of C + 5-MC (methylcytosine, which comes from postreplicative modification of some cytosine residues). The reason is because the geometry of the T and A are such that they will allow for hydrogen bonding, whereas the geometry of the G and C will not allow for hydrogen bonding, which is what accounts for the bonding of two polymeric strands of DNA to form a classic double helix, which composes most DNA. These equal amounts of A and T and C and G in double-stranded DNA is called **Chargaff's rule**, discovered in 1950 by Erwin Chargaff (Chargaff 1950).

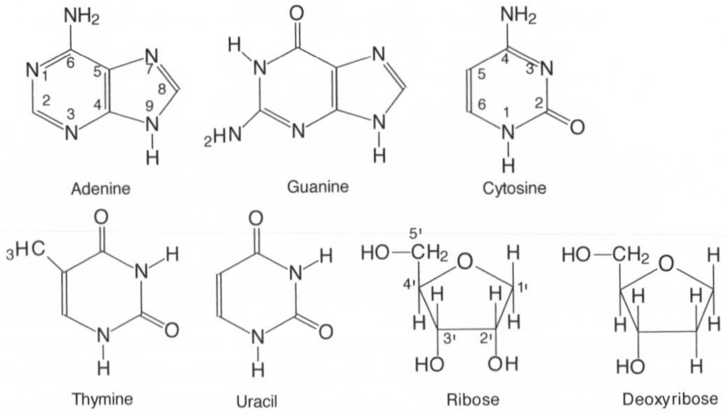

Adenine   Guanine   Cytosine

Thymine   Uracil   Ribose   Deoxyribose

Correct: **nucleoside** – a molecule consisting of purine or pyrimidine base linked to a pentose (a sugar). Adenosine is a nucleoside.

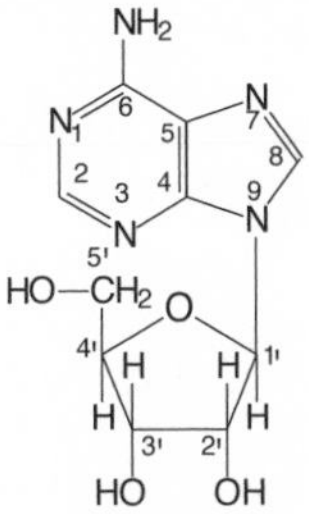

Adenosine
A Nucleoside

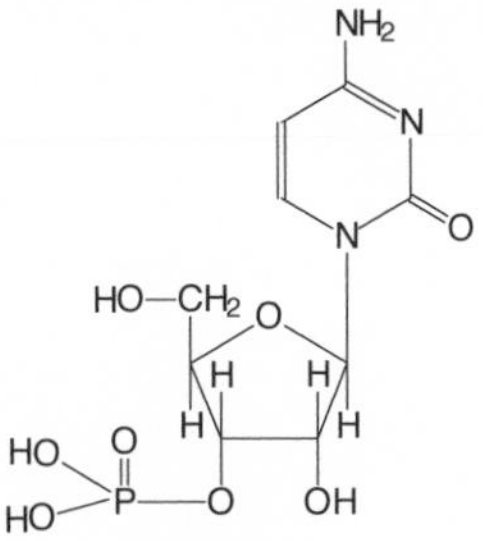

Cytidylic Acid
A Nucleotide

Correct: **nucleotide** – a molecule consisting of a purine or pyrimidine base and pentose with a phosphate moiety. In **ribonucleotides**, the sugar is D-ribose; and in **deoxyribonucleotides**, the sugar is 2′-deoxy-D-ribose. DNA and RNA are polymeric nucleotides (**polynucleotides**), which are also known as nucleic acids.

Correct: **N-terminus** (a.k.a. **amino terminus** or **N-terminal group**) – the residue at the end of a peptide that has an amino group. It may be acylated or formylated (see **C-terminus**).

Correct: **oligomer** – used to describe biological polymers. The strict definition of oligomer found in most English-language dictionaries is "a polymer consisting of two, three, or four monomers". However, the prefix *olig-* is used in biochemistry to have a number of specific meanings (i.e., **oligopeptide**: a polymer of less than 10 amino acids; **oligonucleotide**: a short polymer of 2 to 20 nucleotides; **oligosaccharide**: a carbohydrate consisting of a small number of sugars). Oligomers can be composed of condensations of the same molecule or of different molecules (**heteropolymers**). The term *oligomer* is also used to refer to small proteins that have identical subunits that are called **protomers**. A protomer may consist of one polypeptide chain or several unlike polypeptide chains.

Correct: **peptide** – two or more amino acids linked by a peptide bond that is the condensation of the carboxylic acid group (–COOH) of one molecule with the –NH_2 group of the other molecule. Peptides are condensation polymers of amino acids where the –NH_2 group is on the first carbon (the α-carbon) following the carboxylic acid group. A specific peptide will have a unique sequence of amino acids. When two or more peptides are condensed, the result is a polypeptide.

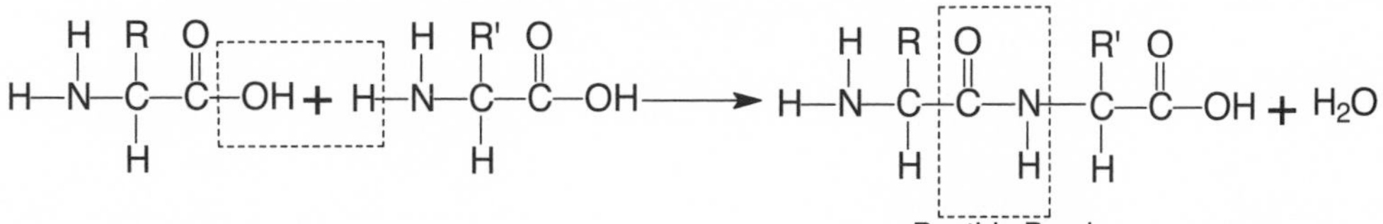

Peptide Bond

Dipeptide

Correct: **pharmacokinetics** (**PK**) – the study of the metabolism and action of drugs. These studies have particular emphasis on the time required for absorption, duration of action, distribution in the body, and excretion. **DMPK** – drug metabolism and pharmacokinetics.

Correct: **polyacrylamide gel electrophoresis** (**PAGE**) – using gels prepared by the free-radical-induced polymerization of acrylamide and N,N′-methylenebisacrylamide in a chosen buffer, proteins can be separated according to their molecular size in solution (the "Stokes radius") by use of an electric field. **Sodium dodecyl sulfate** (**SDS**) will denature the protein's tertiary structure in solution, rendering all proteins in an approximately spherical conformation in which their cross-sectional radius is roughly proportional to their mass. SDS-containing PAGE gels separate proteins in order of their molecular mass with an accuracy of 5–10%. Proteins have charged groups of both polarities; therefore, they have **isoelectric points** (the pH at which the protein is immobile in an electric field). When a protein mixture is electrophoresed through a solution having a stable pH gradient that smoothly increases from the anode to the cathode, each protein will migrate to the position in the gradient that corresponds to its isoelectric point. Two-dimensional polyacrylamide gel electrophoresis (2-D PAGE) (O'Farrell 1975), in which proteins are separated according to charge (pI) by isoelectric focusing (IEF) in the first dimension and according to size (M_r) by SDS-PAGE in the second dimension, has a unique capacity for the resolution of complex mixtures of proteins, permitting the simultaneous analysis of hundreds or even thousands of gene products.

Correct: **polymerases** – a collective term for enzymes that catalyze the formation of macromolecules from simple components (e.g., DNA-polymerase, RNA-polymerase, RNA-synthetase, RNA-dependent, etc.).   **PCR**: polymerase chain reaction; **QPCR**: quantitative polymerase chain reaction; and **RT-PCR**: real-time PCR.

Correct: **posttranslational modification** (of proteins) – the modification of an expressed protein. These modifications can include the covalent bonding of nucleic acids; carboxyl, methyl, acetyl, and/or hydroxyl groups; lipids; carbohydrates; and phosphorus- and sulfur-containing moieties. Formation or reduction of disulfide bonds between cysteine residues is a common posttranslational modification. Some of the terms associated with posttranslational modifications of proteins, which are self-explanatory, are **phosphorylation**, **glycosylation**, **sulfonation**, and so forth.

Correct: **protein** – a polypeptide having a molecular mass between ~6 kDa and several million. Proteins consist of several long-chain polypeptides. A protein's **primary structure** is its sequence of linked amino acids. The coiling or pleating of polypeptides that composes the protein is described as a protein's **secondary structure** (2° structure). Secondary structures are helices, pleated sheets, and turns (coils). The three-dimensional folding of the protein's 2° structural elements along with the spatial disposition of any side chains is the protein's **tertiary structure** (3° structure). The structural relationship of the component polypeptides comprises the protein's **quaternary structure**. When a protein folds so that the –SH portion of two cysteines are adjacent to one another, a disulfide bond can form. Proteins are broadly classified as **globular** or **fibrous**. *Globular proteins* are compact rounded molecules that are usually water soluble. Examples of globular proteins are enzymes (proteins that catalyze biochemical reactions); antibodies (proteins that combine with foreign substances in the body); storage proteins (e.g., caseine in milk and albumin in egg whites); hormones (e.g., insulin); and carrier proteins like hemoglobin. *Fibrous proteins*, which are generally water insoluble, have long coiled strands or flat sheets that result in substances that have strength and elasticity. The primary fibrous proteins of muscle are actin and myosin of which the interaction causes contractions of the muscle. When heated over 50 °C or mixed with strong acids or bases, proteins lose their tertiary structure and can form insoluble coagulates. This denaturing process results in a loss of biological properties (e.g., putting an egg white into boiling water). The tertiary structure can also be denatured by introduction of detergent or surfactant molecules such as sodium dodecyl sulfate or *n*-hexyl glucoside.

The term *protein* (Gk., proteus, L., primarius) was given to the complex organic nitrogenous sustances extracted from plant and animal cells by Geradus Johannes Mulder (1802–1880) in 1839 (Mulder 1839) who had earlier received the suggestion from Jöns Jacob Berzelius (Swedish chemist, 1779–1848).

Correct: **protein database** – an electronic database that is accessible through the Web. These databases are searchable by the names of the protein, their average molecular mass (M_r) and isoelectric point (pI). They are NOT databases of mass spectral data. The University of California San Francisco's Protein Prospector (http://prospector.ucsf.edu/) lists 12 different protein databases: Genpept, SwissPort, Owl, to name a few. Other data-mining sites list TrEMBL, PROSITE, and others. In order to use mass spectral data with these databases to obtain identifications, software tools are necessary. A list of some of the available tools can be found at a Web site maintained by the University of Wisconsin (http://proteomics.mcw.edu/resources/software.jsp). This site was founded in 2002. The software tools are divided into Academic and Commercial. A number of the Academic links are broken. Only the Thermo Electron ProteomeX Workstation and **SEQUEST**™ and MDS Proteometrics (PepSea) links are broken in the commercial category. However, the link to SEQUEST in the Thermo Electron description is still valid for a site maintained by John Yates. One of the more popular software products is **MASCOT**™ from Matrix Science and distributed by several mass spectrometry manufacturers.

Correct: **proteome** and **proteomics** – refer to the proteins that are encoded and expressed by the genome. The terms are generally attributed to Marc Wilkins (executive vice president and head of Bioinformatics, Proteome Systems, Sydney, Australia, and Boston, Massachusetts) at a conference in Siena, Italy, in September 1994. Wilkins defines proteomics as "the study of proteins, how they are modified, when and where they are expressed, how they are involved in the metabolic pathways and how they interact with each other". The first paper using the term *proteome* appeared in 1995 (Wasinger 1995), and the seminal proteome paper also appeared in 1995 (Wilkins 1995). The exact origin of proteomics is not known but appeared ca. 1997, 1998.

Correct: **residue** – the portion of a monomer that remains in a polymer after condensation (e.g., the alanine residue (H_3C–$CH(NH)$–CO–) in a peptide resulting from the formation of a peptide (amide) bond between the –NH_2 portion of alanine (H_3C–$CH(NH_2)$–$COOH$) and another amino acid and alanine's carboxylic acid portion and a third amino acid). In the example of peptide formation, the residue will have a nominal mass 18 Da less than that of the reacted amino acid. Note, the central carbon of an amino acid residue is the α-carbon to which the amine nitrogen, carboxylic acid carbon, a hydrogen, and a **sidechain** group are bound. Alanine's sidechain is a methyl group (H_3C–). Glycine is the only amino acid without a sidechain group.

Correct: **saccharides** – compounds of the general composition $(C \bullet H_2)_n$ where n is ≥ 3. Saccharides are also known as **carbohydrates** because of this general elemental composition. **Sugars** are saccharides. Saccharides can be composed of one or more carbohydrate units that exist as either **hemiacetals** or **hemiketals**. Sucrose (table sugar) is a disaccharide resulting from the condensation of the hemiacetal glucose and the hemiketal fructose.

Correct: **shotgun proteomics** – the method for the rapid analysis of complex protein mixtures by mass spectrometry. This methodology is capable of distinguishing a target species against a large database of background species from a single-component sample or dual-component mixtures with relatively the same concentration. There is an urgent need for rapid detection and positive identification of biological threat agents, as well as microbial species in general, directly from complex environmental samples. This need is most urgent in the area of homeland security but also extends into medical, environmental, and agricultural sciences. Studies have pointed to the potential application of shotgun proteomics for biological threat agent detection; but many areas of research such as sample processing, LC–MS/MS system development, and detailed studies of the effects of database size are needed before this technology will be useful in real-world samples.

Correct: **single-nucleotide polymorphism (SNP)** – a genetic variation pattern in which one or another of two DNA bases is found at an identical genomic location in different people.

SOME OTHER IMPORTANT DEFINITIONS

Note: There are several important words that can create a great deal of confusion in GC/MS, LC/MS, and mass spectrometry because of their double or multiple meanings. The following are the multiple definitions for some of these words:

calibration – mass calibration done by the data system to assign digital-to-analog conversion (DAC) values to mass spectral peaks of some substance that produces ions of known *m/z* values; calibration of the *m/z* scale of the mass spectrum.

– a plot of target analyte amounts versus chromatographic peak area or height, response factor, (etc.); *not* mass calibration.

tuning – adjustment of the mass spectral peak shape, and the resolution between peaks of two adjacent *m/z* values.

– adjustment of the relative intensities of mass spectral peaks during the analysis of compounds such as decafluorotriphenylphosphine (DFTPP) (e.g., those at *m/z* 198 and 442).

peak – **mass spectral peak** – produced by the energy distribution of ions of a given mass-to-charge ratio that strikes the detector of the mass spectrometer.

– **chromatographic peak** – produced by plotting the intensity of signals corresponding to ions of different *m/z* values in consecutive spectra that were acquired during the elution of a compound. These intensities may be directly related to the concentration of the analyte during ionization, depending on whether or not there is coelution.

TIC – In GC/MS and LC/MS, TIC (*total ion chromatogram*) is often used *incorrectly* to describe a reconstructed total-ion-current chromatogram.

– TIC is also used in U.S. EPA methods to describe *tentatively identified compounds*.

– TIC can also be used to describe the *total ion current* in a mass spectrum or display.

library – a collection of mass spectra used to compare against a spectrum of an unknown compound. Mass spectral libraries (a.k.a. databases) are used in mass spectral library searches.

– a database of proteins that is a listing of the amino acids of each protein in the library. These libraries *do not* contain mass spectra.

deconvolution – the separation of coeluting components, or of spectra, in a reconstructed chromatographic peak (chromatographic or spectral deconvolution).

– the determination of the mass of an ion based on the mass spectral peaks that represent multiple-charge ions (charge deconvolution). More appropriately, charge deconvolution should be called **charge-state transformation**.

Correct: **sensitivity** – is actually the slope of a plot of analyte amount versus signal strength.   Unfortunately, the term is *incorrectly* used by many instrument manufacturers to pertain to the amount of analyte put into the primary sample inlet of an instrument that produces an "acceptable" signal (more accurately: the **limit of detection**).   The "sensitivity specification" of an instrument should contain the instrument conditions and a defined result (e.g., in GC/MS, 2 pg of hexachlorobenzene of a sample injected onto a GC column will produce a mass chromatographic peak of m/z 282 with a signal-to-background of 10:1 and a spectrum identifiable as hexachlorobenzene when searched against the NIST/EPA/NIH Mass Spectral Main Library).   Important instrument conditions for a GC/MS sensitivity specification are GC column (length, diameter, stationary phase and its thickness), linear velocity of carrier gas, method of column interface with mass spectrometer (jet separator, open-split interface, direct to ion source, etc.), GC-column oven-temperature program rate, mass spectral acquisition rate, ionization energy, and detector gain (a new detector can be operated at a higher gain to get better sensitivity but will not represent typical operation because detector life is diminished and noise rapidly increases under these conditions).

Care must be taken with respect to the evaluation of a sensitivity specification.   **Figure 12** shows the mass spectra of three different compounds used to demonstrate the sensitivity of a GC-MS.   The percent of ion current represented by the ion used to produce the mass chromatographic peak differs in each case.   It is easy to see that a sensitivity specification of 1 pg of octafluoronaphthalene is equivalent to a sensitivity specification of 2 pg of hexachlorobenzene.

Another factor that has to be taken into account in evaluating a sensitivity specification is the presence of chemical background.   The use of hexachlorobenzene as a detection-limit standard can be complicated by the [X + 1] ion associated with the column-bleed ion at m/z 281.   The m/z-282 isotope from this GC column bleed will significantly contribute to background and will result in a reduction of the signal-to-background value.

The *Current IUPAC Recommendations* defines sensitivity as follows. "Two different measures of sensitivity are recommended.   The first method, which is suitable for relatively involatile materials as well as gases, depends upon the observed change in ion current for a particular amount or change in flow rate of the sample through the ion source.   The recommended unit is a coulomb per microgram (C μg^{-1}).   A second method stating sensitivity that is most suitable for gases depends upon the change of ion current related to the change of partial pressure of the sample in the ion source.   The recommended unit is ampere per pascal (A Pa^{-1})."   However, sensitivity is not defined this way by the instrument manufacturers. The *Current IUPAC Recommendations* does differentiate between sensitivity and detection limit and does specify the importance of the specification containing the experimental conditions.

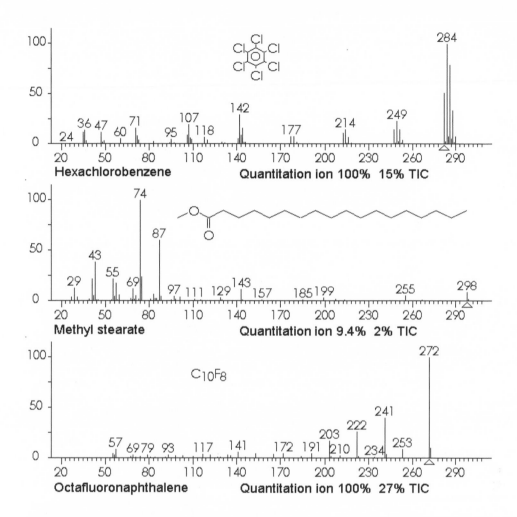

Figure 12. Mass spectra of three compounds that were used to demonstrate the sensitivity in GC/MS instrumentation.

Correct: **signal-to-background ratio (S/B)** – the measure of the signal strength for a sample compared to a region with no sample during a single measurement. S/B is often incorrectly called signal-to-noise. An example of signal-to-background ratio in GC/MS is the measurement of the height of a mass chromatographic peak compared to the height of the baseline, disregarding the zero offset.

Correct: **signal-to-noise ratio (S/N)** – the measure of the noise in the presence of the signal. S/N is the variation of several subsequent measurements of peak height (mass spectral or chromatographic) for the same amount of sample introduced. The variation is the noise, and the peak height is the signal. *Signal-to-background* is a measure of detectability. *Signal-to-noise* is a measure of precision.

Correct: **precision** – the variation within a single measurement (noise) or from measurement to measurement, or the degree of significant figures used to report a number. Precision is related to, but different from, accuracy. *Precision* indicates degree of detail; *accuracy* indicates correctness. The number 3.1416 is more precise than 3.14 but is less accurate for pi.

Incorrect: **precision** – when used to describe the difference between calculated and measured mass.

Correct: **accuracy** – the difference between the actual mass (based on literature values for the individual isotopes) and the measured mass.

Correct: **limit of quantitation (LOQ)** – the smallest amount of an analyte that can be determined accurately and with a set precision. The LOQ should not be confused with the limit of detection.

Correct: **limit of detection (LOD)** – the smallest amount of analyte that will produce a signal that results in an unambiguous confirmation of its presence. The LOD can be below the LOQ, but the LOQ cannot be less than the LOD.

Correct: **ion suppresion** – any process that inhibits ion formation in mass spectrometry. This term is most often encountered in discussions of sensitivity related to electrospray ionization.

Correct: **space charge** – the term used to describe the cause of the broadening of the mass spectral peak in an ion trap when too many ions are present. Because most mass spectral data are presented in a bar-graph form, the effects of space charge are a series of peaks at higher and lower *m/z* values along with peaks for the specific ion. This "Christmas tree"-shaped peak cluster is a result of space charge. Because space charge occurs when the sample pressure is too high, it has been incorrectly referred to as self-CI.

Correct: **ion/molecule reaction** – the term used to describe ALL reactions that take place between ions and molecules in the ion source of a mass spectrometer. In an ion trap mass spectrometer (quadrupole or magnetic), molecular ions of certain compounds undergo a γ-hydrogen shift-induced rearrangement followed by a beta (β) cleavage to produce a distonic odd-electron ion that is a very good proton donor.

These distonic ions react with analyte molecules of the same analyte to produce protonated species because these types of molecules have high proton affinities. These ion/molecule reactions have been referred to as *self-CI*. The term *ion/molecule reaction* describes the process that occurs in chemical ionization mass spectrometry.

Self-CI can also mean the chemical ionization of an analyte where the reagent ion is from an ionized form of the analyte itself (i.e., a fragment ion or a molecular ion of the analyte that protonates other molecules of the same analyte, *or* analyte molecular ions or EI fragment ions that ionize other molecules of the analyte by charge transfer or addition).

Incorrect: **self-CI** – when used to describe space charge and ion/molecule reactions in a quadrupole ion trap mass spectrometer. Self-CI is a term first used by Wilkins and Gross (Ghaderi 1981) to describe the protonation of analyte molecules by fragment ions that result from the electron ionization of the same analyte in an FTICR-MS. Self-CI has been incorrectly used to describe not only this phenomenon that can take place inside an internal ionization quadrupole ion trap (I^2QIT) mass spectrometer or FTICR-MS but also the space-charge effect, which is mass spectral peak broadening (loss of resolution) brought about by having more ions than can be stored in the trap.

Correct: **SI unit** (Système International d'Unités derived from the *m.k.s. system* (meter, kilogram, second), thereby replacing the *c.g.s. system* (centimeter, gram, second) and the *f.p.s. system* (foot, pound, second) for all scientific purposes) – for most units in mass spectrometry.

For example, the use of **pascal** (abbreviated **Pa**: the derived SI unit of pressure, equal to 1 newton [the force required to accelerate a standard kilogram at a rate of 1 m/s^2 over a frictionless surface] per square meter) can be substituted with **torr**[*] (760 Torr is equal to 1 atmosphere [1 atmosphere is the force exerted by the pressure of the atmosphere at 0 °C sea level and 45° latitude on an open reservoir of mercury (density 13.5950 g/cm^3 at 0 °C) so that a column of mercury in a tube with one end sealed and the other end opened and immersed in a reservoir of mercury can be supported to a height of 760 mm. 1 atm = (13.5950 g/cm^3)(980,665 cm/s^2)(76.00 cm)]. This procedure is the origin of the expression *centimeters-of-mercury* or *inches-of-mercury pressure*). Because pressure is a ratio of force to area and not to length, the term "pascal" is more descriptive; however, in the recent past, the term "torr" was more common in mass spectrometry. By definition: 1.013×10^5 Pa = 1 atm = 760 Torr. 1 Pa is $\cong 0.0075$ Torr, and 1 Torr = 133.322 Pa.

A pressure unit that is often encountered on ion-gauge readouts is the **millibar** (**mbar**), which is the standard unit of pressure in the *c.g.s. system*. A bar is a pressure of 10^6 dynes per sq. cm or 10^5 Pa. The mbar = 100 Pa or 0.76 mm Hg. The millibar is the term used by many of the European manufacturers of mass spectrometers before standardization of the vacuum industry on the English system (*f.p.s.*) and the torr. In the *f.p.s.* system, the correct units for pressure are those found in automobile tires in the United States—psi (pounds per square inch).

Another non-SI unit found in the mass spectrometry literature is **kcal mol^{-1}** (mol is the symbol for mole) or **eV** (the energy of an electron accelerated by falling through a potential of 1 volt) as the unit of energy instead of the **joule** (the SI unit for energy; symbol: **J**). The eV molecule^{-1} often appears in older mass spectrometry publications. One calorie is equivalent to $\cong 4.1868$ J. The joule replaced the calorie as the unit of heat in 1948. 1 kJ is equivalent to about 0.24 kcal or 240 calories.

---

[*] Although the torr unit of measure is an abbreviation of the name of Galileo's successor, Evangelista Torricelli (1608–1647), it is treated as the entire name and is written with the first letter as lower case when it is used without a number. As a lowercase abbreviation, torr is the one exception to the rule for the use of proper names and their abbreviations as units of measure when not associated with a number.

Note: When a unit is named after a person, the first letter of the abbreviation is upper case. If the name is spelled out, then the first letter is lower case (i.e., **Da** and **dalton**). The abbreviation should only be used after a number (U.S. Government Printing Office 1986).

Note: The SI system of reporting one unit per another unit (e.g., grams per milliliter) is usually written as "first unit second unit^{-1}" (i.e., µg mL^{-1} rather than µg/mL). Most publications accept either form.

Note: The use of "gms" or "gm" as the symbol or abbreviation for grams or gram, respectively, is considered archaic. The correct symbol for **grams** or **gram** is the lowercase letter **g**.

Although the lowercase letter **l** was used as the symbol for liter for many years, the accepted symbol is an uppercase **L** (e.g., **mL** instead of ml, or **nL** instead of nl). When referring to microliters, do not use the symbol λ (the Greek lowercase lambda)—use *µL*, *not* uL.

A good reference for abbreviations, symbols, and general style information is *The ACS Style Guide: A Manual for Authors and Editors*, J. S. Dodd, Ed., first published in 1986; now in its 2nd edition (1997). Even though different journals and editors have some deviations from *The ACS Style Guide*, it is a good resource. Just as a point-in-fact, in *The ACS Style Guide*, 2nd ed., p 170, Table 6: Non-SI Units That Are Discouraged, the term *torr* appears. This difference between the mass spectrometry standards and the standards of the ACS shows the somewhat rebellious nature of mass spectrometrists in regard to nomenclature and terminology.

Correct: **analyte** – the material (usually an individual compound) in the sample that is being determined by the analysis. However, it should be noted that an analyte can represent a mixture, such as the case of the analysis of soil for gasoline. A capillary electrophoresis purification of a sample results in a pure analyte.

Correct: **absorption** – to take (something) in through or as through pores or interstices. To retain wholly, without reflection or transmission.

Correct: **adsorption** – the accumulation of gases, liquids, or solutes on the surface of a solid or a liquid.

Correct: **system blank** – refers to a run of an analytical method (usually GC or LC, including GC/MS and LC/MS) that involves all aspects of the analysis except the introduction of a sample. An example would be where a GC/MS run was carried out (temperature program started and run, solvent delay applied, the mass spectra acquired at the same rate as would be performed in a analysis after a sample injection, and all electrical settings normally used [EM voltage, ionization energy, etc.]). The resulting data file would be referred to as a system blank data file.

Correct: **analytical blank (AB)** – an analysis of a sample that contains all the components present except for the analyte. An AB can be as simple as the solvent system used to deliver the analyte or as complex as the matrix containing the analyte. In the case where an analysis involves an internal standard, two AB are required—one that contains the internal standard and another sample that does not contain any internal standard.

Correct: **matrix** – the material containing the analyte. This can be from the source of the original sample, a solvent that the sample has been extracted into or dissolved in, a digestant of the original sample, or any material that is not of interest in the analysis being performed (see **sample**).

Correct: **sample** – a mixture of the analyte(s) and a matrix (the material that contains the analyte). The term *sample* can be used to describe the material taken for analysis (i.e., a plasma sample that contains a drug and its metabolites) or the material put into the instrument for the determination of analytes (i.e., the MALDI matrix that contains the analyte).

Correct: **analytical standard** (**AS**) – a sample containing an analyte of known composition. An AS usually consists of a known amount or concentration of the analyte. The AS may be in a solvent system consisting only of a pure solvent, or it may be in the same matrix as the samples to be analyzed against the AS. If the matrix is used for the AS, a determination of the analyte and its amount of endogenous matrix must be made and factored into the analysis.

Correct: **internal standard** (**IS**) – a term used to describe a substance of a known amount and/or composition added to a sample. The purpose of an IS in GC/MS and LC/MS may be to measure the response of the analyte relative to the response of the IS in order to effect a quantitation. Internal standard is also used in mass spectrometry to calibrate the *m/z* in the presence of an analyte ion to allow for an accurate mass determination. Two substances are often used as internal standards—stable isotope-labeled analogs of the analyte and compounds similar to analytes referred to as *surrogate internal standards*.

Correct: **primary standard** – a substance that is unaffected as to composition or purity by normal environmental factors that would be encountered by a human being (i.e., temperatures in the range of 0–45 °C, variations in pressure due to normal changes in altitude, and humidity).

Correct: **selectivity** – refers to the capacity to distinguish or choose. A mass spectrometer can select between ions of different *m/z* values.

Correct: **specificity** – refers to the quality of making selections based on unique properties. A mass spectral analysis can be termed specific when the analyte is characterized by an ion (or group of ions) with a unique *m/z* value not exhibited by any other analyte or substance in the analysis matrix or (in the case of a chromatographic/mass spectral analysis) by a characteristic ion that may be exhibited by another analyte or contaminant but produces a signal that will constitute a chromatographic peak (when the specified ion current is monitored as a function of time) at a specified retention time.

Comments on selectivity and specificity as they relate to mass spectrometry:

If two analytes have the same chromatographic retention time and all the ions in their mass spectra have coincident abundances and *m/z* values, then the mass spectrometer can be *selective* for these two compounds but is *specific* for neither.

Mass spectrometers are inherently *selective* but are only *specific* when constrained by a well-defined experiment.

A gas chromatographic detector that responds only to compounds that contain nitrogen is a specific detector for nitrogen-containing compounds. A compound containing only atoms of carbon, hydrogen, nitrogen, and oxygen will have a specific exact mass based on the specific number of atoms of each element it contains.   However, even though a mass spectrometer can be calibrated to allow only ions of that exact mass to reach its detector, the mass spectrometer is not a specific detector for that compound.

Correct: **derivative** – a compound produced from another compound, retaining the same basic structure and elements as the original compound.   The process of preparing a derivative is **derivatization**.   An analyte is sometimes **derivatized** prior to analysis.   Derivatization is often done to allow for volatility of nonvolatile or thermally labile substances. Derivatization can enhance chromatographic separation.

Incorrect: **effluent** – when used to describe the material that is exiting a chromatographic column.   This word is defined as "something that flows out or forth, especially a stream flowing out of a body of water, an outflow from a sewer or sewage system, or a discharge of liquid waste, such as from a factory or nuclear plant".   The *eluate* from a chromatographic column could be called the column's effluent, but *eluate* is a better choice of words because this word means both the material eluted from the column and the material that accomplishes the elution; whereas, effluent only means an outflow.

Correct: **eluate** – the material that comes from the chromatographic column that consists of the mobile phase, analytes, any dissolved stationary phase, and any components introduced with the sample.

Correct: **eluant** – the mobile phase in a chromatographic process; the substance used in the process of elution.   The word *eluant* has also been spelled **eluent**, which is a modern word that first appeared during the World War II era (1940–1945).

Correct: **good laboratory practice** (**GLP**) – a set of procedures to ensure that the quality of data is high and the results can be used for various purposes. GLP has been adopted by many different types of laboratories and is essential in the pharmaceutical industry.   An important aspect of GLP is that all necessary information is recorded to reacquire and reprocess a sample and is saved with the data file.   Other acronyms used are **GALP** (good automated laboratory practice) and **cGLP** (current GLP).

Correct: **intramolecular** – involves only the decomposition of individual ions.

Correct: **intermolecular** – involves the reaction of ions with other ions or neutral molecules or fragments.

Correct: **mean-free path** – the minimum length that half of all ions present in an *m/z* analyzer will travel before they encounter another ion or molecule.

Correct: **silylation** – a process of reacting an analyte with a reagent to produce a silyl derivative.   Trimethyl silyl derivatives are often made of sugars to aid in their volatility for analysis by electron ionization mass spectrometry. Silyl derivatives can be produced in order of reactivity from aliphatic alcohols, phenols, thiols, carboxylic acids, *primary* and *secondary* amines, and amides, with the reactivity within aliphatic alcohols being primary > secondary > tertiary.

ISOMER NOMENCLATURE

Organic chemistry (and therefore organic mass spectrometry) deals with isomers of molecular formula. Isomers are different compounds with the same molecular formula. This section provides some of the nomenclature that will be encountered when dealing with isomers. Very good references for a more comprehensive treatment of the definition of stereochemical terms are "Glossary" in Eliel, EL; Wilen, SH (with a chapter on stereoselective synthesis by Mander, LN) *Stereochemistry of Organic Compounds*; Wiley: New York, 1994; and Smith, MB; March, J *March's Advanced Organic Chemistry: Reactions, Mechanisms, and Structure*; Wiley–Interscience: New York, 2001, or earlier editions.

Correct: **achiral molecule** – a molecule whose mirror images are superimposable on one another. If a molecule has a plane of symmetry, it is **achiral** (**nonchiral**).

Correct: **configuration** – refers to isomers that can be separated by physical means.

Correct: **conformation** – as related to organic compounds, refers to three-dimensional arrangements in-space of the atoms in a molecule that are interconnected by a free rotation about single bonds. Conformations are rapidly convertible, making them inseparable. Individual conformational isomers are called **conformers** and **rotamers**.

Note: Conformations are illustrated using either **Newman projection** or **sawhorse formula**. Conformers (rotamers, conformational isomers) exist in **eclipsed** or **staggered** conformations. The resistance to rotation about the single bond is the **torsional barrier** of the single bond. The staggered conformation is more stable than the eclipsed.

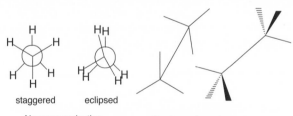

staggered   eclipsed

Newman projection   Sawhorse formula

The above illustrations are representations of ethane. In the Newman projection, the lines to the circle represent the attachments to the rear atom; and those to the center represent attachments to the front atom. When the molecule is more complex such as butane and the rotation of the C2–C3 bond is considered, the staggered conformation can have an *anti* and *gauche* form as illustrated below.

anti conformation   gauche conformations

Cyclohexane systems have **boat** (usually higher-energy form) and **chair** conformations. The chair conformations have *axial* and *equatorial* positions. The **axial bonds** can point above or below the plane of the ring. The **equatorial hydrogens** lie on the perimeter of the ring pointing in a nonperpendicular direction away from the ring. In naming a substituted cyclohexane, the bond position should be specified (i.e., equatorial *tert*-butylcyclohexane or axial *tert*-butylcyclohexane).

Correct: **congener** – a compound that is similar. The series H_3CCl, H_2CCl_2, $HCCl_3$, and CCl_4 are congeners. The three constitutional isomers of xylene are congeners.

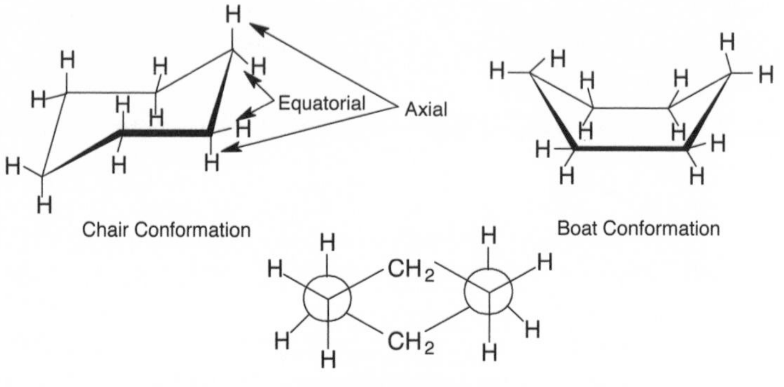

Chair Conformation

Boat Conformation

Newman projection of the chair conformation of cyclohexane

Correct: ***cis- trans-* designation** – a nomenclature used in the naming of disubstituted alkene diastereomers where the substitutions are on the carbon atoms on either side of the double bond due to the lack of free rotation about the carbon-to-carbon double bond. If the two hydrogens are on the same side of the molecule, it is *cis-*. If the two hydrogens are on opposite sides, the molecule is *trans-*.

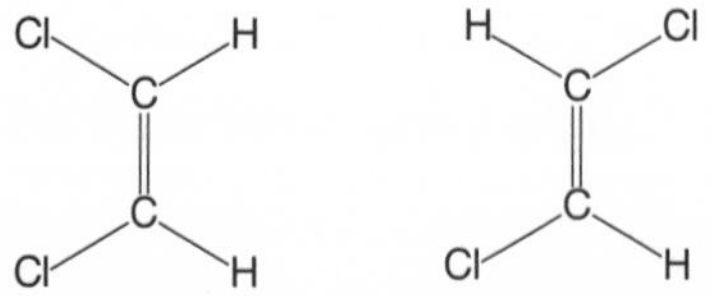

cis-1,2-dichloroethene
($C_2H_2Cl_2$)

trans-1,2-dichloroethene
($C_2H_2Cl_2$)

Incorrect: **chiral atom** or **asymmetric atom** – when referring to a tetrahedral atom with four different substituents. This terminology was discontinued after K. Mislow and J. Siegel explained that the use of these terms resulted in conceptual confusion. Chirality is a geometric property of the chiral molecule. They pointed out that all the atoms of a chiral molecule are in the chiral environment, and all the atoms are said to be *chirotopic*. Consideration of the C2 atom of 2-butanol is with respect to a stereocenter, and this atom should not be designated as a chiral atom but as a *stereocenter* (Mislow 1984).

Correct: **chiral molecule** – a molecule that is not identical (superimposable) with its mirror image. Chiral molecules have at least one tetrahedral atom with four different substituents. This atom is a **stereocenter**.

Correct: **constitutional isomers** – different compounds having the same molecular formula even though the atoms are connected in different orders (e.g., acetone with a structural formula of CH_3COCH_3 and propanol with a structural formula of CH_3CH_2CHO are constitutional isomers as are *n*-butane and isobutane).

Constitutional isomers of disubstituted benzene compounds that differ by the relation of one substituent to another are referred to as **ortho-** [*o*-] (1,2-substitution), **meta-** [*m*-] (1,3-substitution), and **para-** [*p*-] (1,4-substitution). Another term used with respect to substitutions on aromatic rings is **ipso-**. The ipso- position is the position that has a non-hydrogen attached. The term is associated with ipso-attach, which is the attachment of an entering group to the position that has the leaving group in a substitution reaction. The use of ipso-substitution is synonymous with substitution and therefore is not used. The term **cine-substitution** means the entering group takes a position ortho- to the position of the leaving group; and the term **tele-substitution** means that the entering group takes a position more than one atom away from the leaving group.

o-xylene

m-xylene

p-xylene

Correct: **diastereomers** – stereoisomers that are not mirror images of one another. An example would be *cis*- and *trans*-1,2-dichloroethene or (*Z*)- and (*E*)-2-bromo-1chloro-1-fluorethene. Stereoisomers with more than one stereocenter that are not mirror images are also diastereomers.

Correct: **D- L- designation** – a method of designating monosaccharides and α-amino acids using (+)- or (–)-glyceraldehyde as a reference compound. (+)-glyceraldehyde is defined as D- with the OH group attached to C2 on the right side of the Fischer projection formula in which the CHO group appears at the top. The enantiomer, with the OH on the left, is defined as the L-isomer. In carbohydrates, the position of the OH attached to the highest numbered stereocenter C atom determines whether the compound is D- or L-. For α-amino acids, L-compounds are those in which the NH_2 is on the left side of the Fischer projection formula where the COOH group is at the top. D- or L- do not relate to the sign of rotation of plane-polarized light, which is designated by (+)- and (–)- (formerly (*d*- and *l*-). The *d* (dextrorotatory) and *l* (levorotatory) designation regarding the rotation of the sodium D line emission (589 nm) is replaced by the (+)- and (–)- symbolism. The use of *dl* for racemates should be replaced by (±) or *rac*.

Correct: **enantiomers** – stereoisomers that are non-superimposable mirror images of each other. Enantiomers do not differ from one another in their density, vapor pressure, infrared spectra, melting and boiling points, index of refraction, solubility in common solvents, or in their rates of reaction with achiral solvents. They do have different rates of reaction with other chiral substances and different solubilities in solvents that consist of a single enantiomer or an excess of a single enantiomer. They also rotate plane-polarized light equally but in opposite directions as it passes through them. A sample with only a single enantiomer is **enantiomerically pure** or has a 100% **enantiomeric excess**.

Correct: *endo-* and *exo-* – prefixes used to indicate the relation of a substituent to the bridgehead of a bicyclo compound. If the substituent is *cis-* to the bridgehead, then the configuration is *endo-*; if the substituent is *trans-* to the bridgehead, then the configuration is *exo-*.

Correct: **epimers** – two diastereomers that have different configurations at only one chiral center. Epimer originally applied to aldoses of opposite configuration at the number 2 carbon (i.e., glucose and mannose). The term has now been generalized.

Correct: **(*E*)- (*Z*)- designation** – a nomenclature system used in the naming of alkenes that are tri- or tetra-substituted to avoid ambiguity. The (*Z*)-isomer (*zusammen*, German, together) has the two highest priority atoms on the same side of the double bond, whereas the (*E*)-isomer (*estgegen*, German, opposite) has the highest priority atom of the two atoms attached to each of the carbons on either side of the double bond on the opposite sides. The same priority rules are used as for the (*R–S*) system of naming enantiomers.

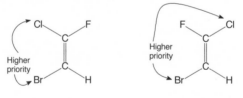

(*Z*)-2-bromo-1-chloro-1-fluorethene   (*E*)-2-bromo-1-chloro-1-fluorethene

Correct: **geminal (*gem*)** – an adjective that means substituents are on the same carbon atom.

Correct: **isomer** – a different molecule, ion, or radical that has the same elemental composition. Because there are so many different types of isomers, the word "isomer" should only be used with a descriptive adjective (i.e., constitutional isomer, stereoisomer, *cis- trans-* isomer, etc.).

Correct: **homologous series** – a group of compounds that are similar but differ by a single unit. A series of straight-chain hydrocarbons C3 to C10 is a homologous series with each member differing by a single methylene unit ($-CH_2-$).

Correct: **meso compounds** – structures with two or more stereocenters that have a plane of symmetry are achiral. An example is cyclic stereoisomers that are diastereoisomers rather than enantiomers. 1,2-dimethylcyclopentane exists as three stereoisomers: *cis*-1,2-dimethylcyclopentane is a meso compound because it has a plane of symmetry even though there are two stereocenters, whereas *trans*-1,2-dimethylcyclopentane does not have this plane of symmetry and exists as a pair of enantiomers. Meso compounds do not rotate plane-polarized light.

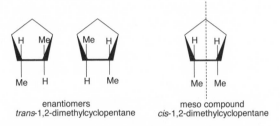

enantiomers
trans-1,2-dimethylcyclopentane

meso compound
cis-1,2-dimethylcyclopentane

Correct: **optical activity** – a term applied to enantiomer rotation of plane-polarized light. Optical activity is sometimes incorporated into the name of a compound by the use of (+) for rotation in a clockwise direction (**dextrorotatory**, *d*); and (–) for rotation in a counterclockwise direction (**levorotatory**, *l*). An old form of designation using *d* and *l* was replaced by the (+) and (–) designation. Whereas only *R* and *S* enantiomers exhibit optical activity, *R* isomers can have a (+) optical activity or a (–) optical activity. *R* and *S* are *not* related to the direction of rotation of plane-polarized light. A good example of the differences in the properties of optical activity of (+) and (–) isomers is the fact that (*S*)-(+)-Carvone is the principal component of caraway seed oil, and (*R*)-(–)-Carvone is the main component of spearmint oil.

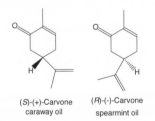

(*S*)-(+)-Carvone
caraway oil

(*R*)-(-)-Carvone
spearmint oil

In naming optically active compounds, the **specific rotation** is sometimes reported as a part of the name; that is,

$$[\alpha]_{25}^{D} = +5.17^{o}$$

states that the specific rotation $[\alpha] = \alpha(c - l)^{-1}$ (where α is the observed rotation) is 5.17° clockwise when measured at a temperature of 25 °C using the D line of a sodium lamp ($\lambda = 589.6$ nm) of a sample containing 1 g mL^{-1} (C, concentration) of the optically active compound in a 1-dm (path length in decimeters) tube.

Correct: **racemic mixture** or **racemate** – an equimolar mixture of enantiomers. The racemate will show no rotation of plane-polarized light.

Correct: **regioisomers** – isomers that have the same functional groups, but these groups are located at different positions (e.g., *n*-propyl alcohol and isopropyl alcohol or *m*-xylene and *p*-xylene).

Correct: **(*R*–*S*) system of naming enantiomers** – (1) a *priority* or *preference* (*a, b, c,* or *d*) is assigned to each of the four groups attached to the stereocenter (or to the two carbons on either side of the double bond in the *E–Z* system). The group (atom) with the lowest atomic number is assigned the lowest priority (*d*). In the case of isotopes, the isotope of the highest mass is assigned the highest priority. (2) When it is not possible to assign a priority based on the atoms attached to the stereocenter, the next sets of atoms in the unassigned groups are examined to give these groups a priority. An ethyl group has a higher priority than a methyl group. Priority is assigned at the first point of difference. Branched chains require that the chain with the highest priority atom must be followed. **Note:** Groups with double or triple bonds are assigned priorities as if atoms are duplicated or triplicated, respectively. (3) The formula is rotated so that the lowest priority atom is facing away from the view. If the trace of the path from *a* to *b* to *c* is clockwise, then the designation is (*R*). If the direction is counterclockwise, then the enantiomers designation is (*S*). *R* (*rectus*, Latin, right) and *S* (*sinister*, Latin, left).

Correct: **stereoisomers** – compounds that have the same connectivity of atoms but differ in the spatial arrangements of their atoms. There are stereoisomers that do not exhibit chirality (diastereomers), and those that do (enantiomers); that is, those that do not and those that do have a mirror-image relationship, respectively.

Incorrect: **structural isomers** – when used to describe constitutional isomers, regioisomers, and diastereomers. The use of structural isomerism was discontinued because it covered too many different types of isomers.

Correct: ***syn- anti-* isomers** – when two different groups are associated on either side of a C=N, N=N, or C=S. If the two non-hydrogen groups on either side of the double bond are the same and are on the same side, they are *syn-*; if they are on opposite sides, they are *anti-*.

syn-   anti-

Correct: **syn–anti dichotomy** – the fact that the *trans-* isomer is formed when the "syn" elimination occurs, and the *cis-* isomer is formed by an "anti" elimination. In an E2 mechanism (elimination, bimolecular) where an adjacent proton is removed by the action of a base at the same time an X radical leaves, if the hydrogen and the group representing the radical or a dihedral angle to one another (180°), they are *trans* to one another and the configuration is **antiperiplanar** (*ap*). If the dihedral angle is 0°, they are *cis* to one another and the configuration is called **synperiplanar** (*sp*). The initial relation of the leaving proton to the leaving radical determines the configuration of the resulting product. Syn–anti dichotomy is a property primarily of ring systems having 8–12 members. It has also been observed in open-chain systems to a lesser extent.

Correct: **vicinal (*vic*)** – an adjective that means substituents are on adjacent carbon atoms.

Correct: **tautomerism** or **dynamic isomerism** – a special case where two constitutional isomers are directly interconvertible as is the case with ketones and enols or tropylium and benzyl ions. The two structures are called **tautomers** because they differ from one another by the location of a H and a pair of π electrons.

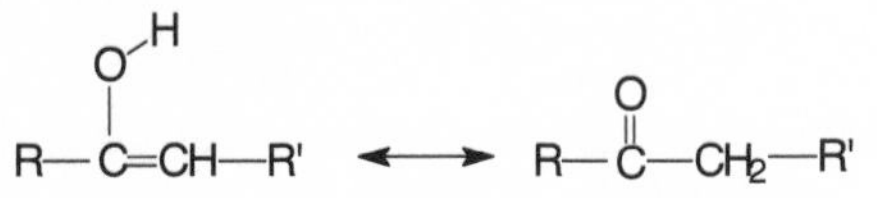

Enol   Ketone

Tautomers

SOME OTHER TERMINOLOGY TO AVOID

Inappropriate: **moiety** – a word that crept into the mass spectrometry literature of the 1960s from organic chemistry. The publication of many mechanistic rationalizations prompted a search for synonyms of fragment. As used in organic chemistry, moiety usually means portion (i.e., the benzoyl portion of the molecule is the benzoyl fragment or moiety). However, as can be seen from the definition below, its primary meaning is "a half". Maurice M. Bursey (Bursey 1991) suggests that this word is not a good English word and should not be used.

> moiety (moi'î-tê) noun
> plural, moieties
>   1.   a half
>   2.   a part, portion, or share
>   3.   either of two basic units in cultural anthropology that makes up a tribe on the basis of unilateral descent
> [Middle English *moite*; from Old French *meitiet, moitie*; from Late Latin *medietâs*; from Latin *medius*, middle]

In spite of Bursey's comments regarding this word, it is still extensively used, and its meaning is readily recognized. Perhaps moiety is one of those words that gives elegance to the language of chemistry.

Correct: **fragment**, **portion**, or **part** – terms used to refer to a portion of a molecule in mass spectrometry (i.e., a benzoyl fragment or loss of the methyl portion of the molecular ion).

Correct: **formula weight** (**mass**) – a term that is generally not used in mass spectrometry. This mass is based on the integer or 0.5 value of an atomic mass (weighted average).

Correct: **quantitate** (verb), **quantitative** (adjective; i.e., *quantitative analysis*), **quantitation** (noun) – terms used to describe experiments that result in an expressed number of units (i.e., grams, meters, g mL^{-1}, etc.). To quantitate is to measure. The word *quantitate* had its origin between 1955 and 1960.

Inappropriate: **quantify** (verb), **quantifiable** (adjective; i.e., *quantifiable analysis* is very different from *quantitative analysis*), **quantification** (noun) – when used to describe an experiment that results in an expressed number of units. To quantify is to give quantity to something (i.e., a *large* amount of money, a *big* man, etc.).

Note: The terms *quantitate* and *quantify* both relate to quantity. The word *quantity* is defined as a specified or an indefinite number or amount; or an indefinite or aggregate amount (i.e., "a quantity of 10 grams" or "a quantity of sugar"). However, *quantitate* has one and only one meaning: "to determine the quantity of, especially with a precision" or "to determine or measure the quantity of". Implicit in either of these two statements defining the term *quantitate* is that the result of an act of quantitation is a specified number.

The term *quantify* has multiple definitions. One of these definitions (to determine, indicate, or express the quantity of) could be interpreted to include the "measurement of". However, the other definitions of quantify are involved with proprieties other than a specific number. These definitions include the use in the logic to make explicit the quantity of (i.e., a proposition) or to give quantity to something that is regarded to have only quality (i.e., "great beauty" or "of little significance"—the questionable use of hyperbole). The origin of these uses of the term *quantify* was between 1830 and 1840.

Another indicator of the inappropriateness of the terms *quantification* and *quantify* with respect to scientific measurements is the derived word *quantifier*, defined as an expression as "all" or "some" that indicates the quantity of a proposition; and a word or phrase that modifies a noun that indicates quantity (i.e., "few" or "a lot").

Quantify implies a degree of subjectivity, and quantitate implies objectivity. Because use of more than one word to indicate the same act, object, (etc.) will lead to confusion, the more exacting of the two terms (quantitate) should be employed. Quantify should be considered archaic with respect to scientific analyses.

Usage: **analysis**, **determination**, and **measurement** – care must be taken in the use of these three words. They are often incorrectly used interchangeably. An **analysis** is the experimental examination for the purpose of measuring the absence/presence or quantity of some or all specified components (e.g., the water sample is *analyzed* for the quantity of pesticides present). Based on such an analysis, it is possible to reach a conclusion as to the quality of the water (e.g., it was *determined* that the water quality is bad based on the *analysis* of the water). "The water quality is *determined* by an *analysis*, which involves the **measurement** of the amount of specific pesticides present." Contents are *measured*; samples are *analyzed*; and **determination** is a result of some process or a conclusion based on some information.

The distinction between analysis and determination was made by F. C. Strong in 1975 (Strong 1975) and reported in several issues of *American Laboratory* in 2001 (Strong 2001-1). The implied definition of the word *determination* made by Strong was challenged by E. D. Stevens (Stevens 2001) who said that the correct term reporting the presence, absence, or quantity of a substance in a sample is measured rather than determined because the word *determination* related to cause rather than quantity or quality measurements. Strong countered, with a statement from the *American Heritage Dictionary* that defined *determination* as "to establish or ascertain definitely, as after consideration, investigation, or calculation", saying, "The investigation, in this case, is the chemical analysis of the sample that establishes the quantity or proportion of the component under consideration in the sample" (Strong 2001-2). However, after further consideration, it appears that it is best to reserve the use of the word *determination* to describe the establishment of cause. This is more in accordance with the other dictionary definitions of determination.

Note: There is a tendency in scientific writing to avoid the use of personal pronouns. There is nothing wrong with the use of personal pronouns in technical presentations. Some authorities encourage the use of personal pronouns because it gets the reader more involved. The word "one" is often used as a poor (and awkward) substitute for the personal pronoun and should be avoided.

Awkward: If one uses gas chromatography/mass spectrometry in the analysis of underivatized amino acids, then one will get little information.

Correct: If you use GC/MS in the analysis of underivatized amino acids, then you will get little information.

Preferred: If GC/MS is used in the analysis of underivatized amino acids, then little information will be obtained (*in the event you do not want to use personal pronouns*).

Either use personal pronouns or don't use them. Do not substitute the word "one" for a personal pronoun.

In lecturing, there will be a tendency to use the term "note that". The inclusion of "note that" can be overused in writing, and its excessive use should be avoided. The same is true for "Thus," to begin a sentence.

INSTRUMENTS

Types of *m/z* Analyzers

double-focusing mass spectrometer: This instrument has a magnetic sector (single-letter symbol **B**) and an electric sector (single-letter symbol **E** a.k.a. electrostatic sector). This type of instrument is often referred to as a **high-resolution** mass spectrometer because of its ability to separate ions that have very small differences in *m/z* values (0.0001). The electric sector is used to produce a dispersion of ions based on their kinetic energy, whereas the magnetic sector produces a dispersion of ions based on their momentum. In a **forward-geometry** instrument (ions pass through the electric sector before the magnetic sector), ions of a narrow kinetic energy pass through a slit to the magnetic sector where ions with a small difference in mass are separated based on their momentum. In a **reverse-geometry** instrument (magnetic sector preceding the electric sector), ions of a common momentum enter the electric sector where they are dispersed based on their differences in kinetic energy. Only those of a narrow kinetic energy are allowed to pass through a slit into the ion detector, thereby accomplishing the precise mass measurement. Sometimes these instruments are described based on the location of the two fields relative to the ion source and the detector (see **Terms Associated with Double-Focusing Mass Spectrometers**) or according to the names of the developers of their particular ion-optic geometry (i.e., Mattauch–Herzog or Nier–Johnson). The limiting factor for the instrument's upper *m/z* limit is the maximum strength of the magnetic field.

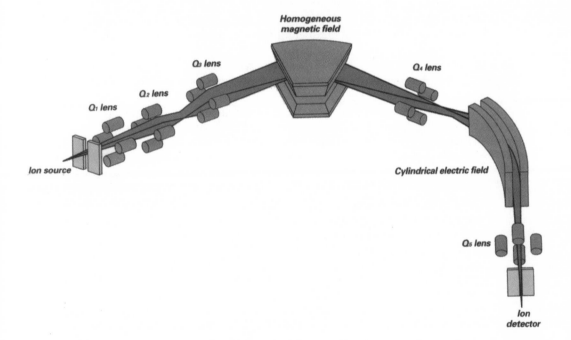

Figure 13. Schematic representation of a reverse-geometry double-focusing mass spectrometer. (Courtesy of JEOL USA, Inc.)

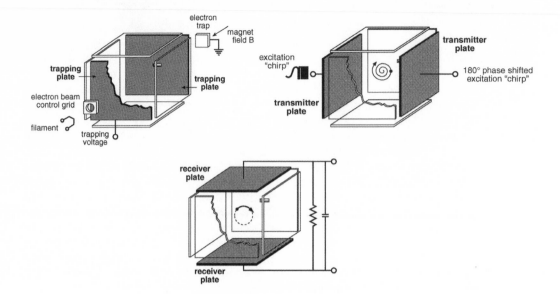

Figure 14. Schematic representation of the ion trap of a Fourier transform ion cyclotron resonance (FTICR) mass spectrometer illustrating the three functions required to obtain a mass spectrum. (Watson, JT *Introduction to Mass Spectromety*, 2nd ed.; Raven: New York, 1997; p 101)

Fourier transform ion cyclotron resonance (FTICR) mass spectrometer: This instrument is a magnetic ion trap (a.k.a. **Penning ion trap**). The technique of Fourier transform ion cyclotron resonance mass spectrometry (FTICRMS) in commercially produced instruments is currently limited to the use of **ion cyclotron resonance (ICR)** mass spectrometers. Some FT ion detection has been experimented within quadrupole ion traps (Badman 1998). In FTICRMS, ions are trapped in a cell (which is inside a strong magnetic field: **B**) composed of three distinct sets of plates: trapping, transmitter, and receiving plates. The ions in the trap move in circular orbits (cyclotron motion) in a plane perpendicular to the magnetic field. The ions are held in the cell by electric potentials applied to the trapping plates that are perpendicular to the magnetic field. Mass analysis is accomplished by the application of a radio frequency electric potential to the transmitter plates to cause trapped ions to be excited into larger circular orbits. As the excited ions pass near the receiver plates, the frequency of their passage is detected as an induced current in the plates called "image" current. The frequency of the motion of an ion is inversely proportional to its mass. This type of ion detection results in a nondestructive detection of ions. The signal-to-background is enhanced by averaging many cycles before transforming and storing data. It is possible to use lasers to fragment ions in these traps to perform MS/MS. These instruments are capable of extremely high resolving power. In order to obtain the desired magnetic field strength, it is necessary to use a superconducting magnet.

magnetic-sector mass spectrometer: This single-focusing instrument has only a single magnetic sector. In modern instruments, ions are accelerated from the ion source at a fixed value of 1–10 kV into a tube of a fixed radius of curvature that is subjected to a magnetic field with the direction of magnetic flux at a right angle to the path of the ions. Under these conditions, only ions of a single narrow *m/z* range will have the same radius of curvature as the tube and reach the end of the tube where a detector is mounted. It is possible to measure all ions of different *m/z* values by incrementally changing the field strength of the magnet or by allowing the ions to impinge on a plane (as opposed to a point) in an instrument with a fixed-field magnet. In this latter case, ions of differing *m/z* values have different radii of curvature and strike the plate in different locations, according to their different *m/z* values. It is also possible to use a fixed-field magnet and incrementally change the accelerating voltage to obtain a mass spectrum. This technique is often employed in selected-ion monitoring with sector-based instruments. These instruments cannot produce much better than unit to 0.1 *m/z* resolution. The equation used to describe ion separation in a magnetic-sector instrument is

$$m/z = R^2 B^2 / 2V$$

where *R* is the radius of curvature of the ion tube, *B* is the magnetic flux density (not *H*, the magnetic field intensity—the magnet's flux density divided by the permeability, *B*/μ), and *V* is the accelerating potential of the ions.

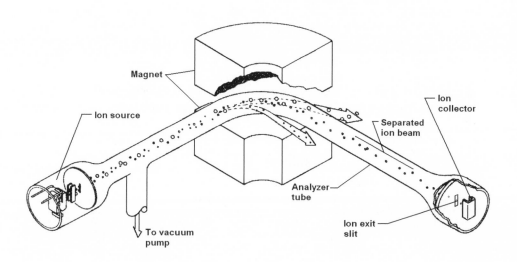

Figure 15. Schematic representation of a magnetic-sector mass spectrometer.
(Adapted from McLafferty, FW; Tureček, F *Interpretation of Mass Spectra,* 4th ed.; University Science Books: Mill Valley, CA, 1993; p 8)

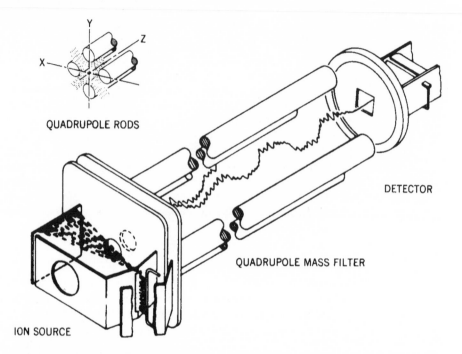

QUADRUPOLE RODS

DETECTOR

QUADRUPOLE MASS FILTER

ION SOURCE

Figure 16. Schematic representation of a transmission quadrupole mass spectrometer with an electron ionization source showing the trajectory of a stable ion. (Watson, JT *Introduction to Mass Spectromety*, 3rd ed.; Lippincott-Raven: Philadelphia-New York, 1997; p 75)

quadrupole or **transmission quadrupole** (single-letter symbol **Q**) mass spectrometer: This instrument separates ions based on oscillations in an electric field (the quadrupole field) created with the use of radio frequency (RF) and direct current (DC) voltages. For a given amplitude of a fixed ratio of RF (at a fixed frequency) to DC, ions of only a single m/z value will oscillate and pass from one end of a set of four poles positioned to form a hyperbolic cross section. All other ions are filtered out of the ion beam. This transmission of ions of a single integer m/z value is the reason why the term **mass filter** is sometimes used when describing this mass spectrometer. A mass spectrum is obtained by increasing the amplitude of the RF and DC while holding their ratio constant. These instruments operate between unit and 0.1–0.3 m/z resolution. The use of "transmission" as an adjective for this type of instrument is to avoid confusion with the other type of mass spectrometer that uses a quadrupole field for the separation of ions—the quadrupole ion trap mass spectrometer (QIT-MS).

triple-quadrupole (QqQ) mass spectrometer: a tandem-in-space instrument used for MS/MS. This term is misleading in that it implies that there are three transmission quadrupole mass spectrometers in tandem. There are only two mass spectrometers (Q_1 or MS_1 and Q_3 or MS_2) separated by a collision cell (q_2) (see **MS/MS** for a definition of tandem-in-space mass spectrometry). This instrument was developed at Michigan State University by Rick Yost and Chris Enke based on discussions with Jim Morrison's group at La Trobe University in Bundoora, Victoria, Australia (Yost 1978).

quadrupole ion trap or **ion trap** (symbol **QIT**) mass spectrometer (a.k.a. **Paul ion trap**): This type of mass spectrometer uses a quadrupole field to separate ions. The use of "quadrupole" as an adjective for this type of instrument is to avoid confusion with the other type of ion trap mass spectrometer—the FTICR-MS. There are now 3-dimensional and linear QIT mass spectrometers commercially available. The 3-D QIT mass spectrometer is also referred to as a *Quistor* (quadrupole ion storage) mass spectrometer.

There are two specific types of 3-dimensional quadrupole ion trap mass spectrometers: (1) instruments where the primary ionization takes place inside the place where ions are stored—the **internal ionization quadrupole ion trap** (**I^2QIT**); (2) instruments where the primary ionization takes place outside the place where ions are stored—**external ionization quadrupole ion trap** (**ExIonQIT**). LC/MS requires the use of an ExIonQIT-MS. GC/MS can be carried out with an ExIonQIT-MS or an I^2QIT-MS.

The QIT mass spectrometer uses a three-dimensional quadrupole field for ion separation. Ions of a range of *m/z* values (based on the frequency of the RF voltage) are stored in a field that is created with a fixed-frequency RF applied to a cylinder ring electrode. End caps held at ground, or subjected to various waveforms at different-frequency RF or DC voltages, are positioned on either side of the ring electrode but are electrically isolated. As the amplitude of this fixed-frequency RF is increased, the trajectories of ions of increasing *m/z* values will become sequentially unstable and move toward the two end caps. Those ions reaching the end cap (or caps) with holes to allow ions to pass to a detector are recorded. The increasing instability of the ions is observed as an increased amplitude of the orbital path of the ion. Another unique feature of the 3-D QIT is its operating pressure of 10^{-1} Pa. This pressure is a result of He used as **bath**, **buffer**, or **cooling gas** to help ions of a given *m/z* value maintain discrete orbitals. The high-pressure He is also used as a collision gas for MS/MS experiments.

The application of various waveforms to the end caps can result in the selected storage of precursor ions for MS/MS or ions specific to an analyte with the exclusion of matrix ions that could result in reduced detection limits. Specific ion storage also allows for very low pressure chemical ionization (10^{-3} Pa for the reagent gas as opposed to 100 Pa in a beam-type instrument) in the internal ionization QIT. Commercial models of QIT mass spectrometers operate at unit to 0.1–0.3 *m/z* resolution; however, a resolving power of 10^7 has been demonstrated (March 1995).

In the first few years of the 21st century, there have been several variations of the quadrupole ion trap (Cooks 2003), which included two commercialized variations of the **linear quadrupole ion trap** (**LIT**) and a cylindrical ion trap, multiplexed mass spectrometers, an orbitrap mass spectrometer, and a rectilinear ion trap. This latter group has no implementations that have been commercialized. Linear ion traps can be used either in a transmission quadrupole mode or as a quadrupole ion storage device with subsequent individual *m/z* destabilization to obtain a mass spectrum. The linear QITs use the spacers that separate the rods that are often used for the electrical poles in a transmission quadrupole as the end caps for the trap mode of the device. One of these systems involves axial ejection of ions (Hager 2002), and the other involves radial ejection of ions through the spaces between the rods

(Schwartz 2002) with dual detectors. In addition to being used as a part of a triple-quadrupole instrument, there have been standalone devices and linear quadrupole ion traps that have been used in tandem with Fourier transform ion cyclotron resonance mass spectrometers.

Resonance electron capture ionization can only be carried out with the ExlonQIT. This ion trap can be used for multiple iterations of MS/MS; however, it will allow only for product-ion analyses. Precursor-ion and neutral loss MS/MS cannot be carried out with the QIT.

Both the transmission quadrupole and the quadrupole ion trap mass spectrometer rely on the amplitude of a fixed-frequency RF voltage to obtain a mass spectrum. The maximum *m/z* limit for these instruments is based on the maximum amplitude that can be obtained without a disintegration of the RF's sign wave.

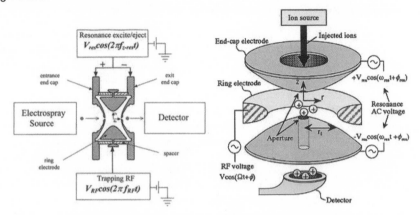

Figure 17. Schematic representations of the 3-dimensional quadrupole ion trap mass spectrometer. (Adapted from Yoshinari, K "Theoretical and Numerical Analysis of the Behavior of Ions Injected into a Quadrupole Ion Trap Mass Spectrometer" *Rapid Commun. Mass Spectrom.* **2000**, *14*, 215–233)

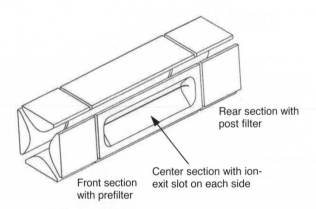

Figure 18. Schematic representation of the linear quadrupole ion trap mass spectrometer with radial ion ejection. (Schwartz, JC; Senko, MW; Syka, JEP "A Two-dimensional Quadrupole Ion Trap Mass Spectrometer" *J. Am. Soc. Mass Spectrom.* **2002**, *13*, 659–669)

Orbitrap mass spectrometer: This is the latest in trapping devices used as an *m/z* analyzer. This device was developed by Alexandor Makarov while at HD Technologies, Ltd. in the United Kingdom (Makarov 1999, 2000). The Orbitrap has now been commercialized as part of a tandem-in-space mass spectrometer using a linear quadrupole ion trap developed by Thermo Electron (San Jose, California) and introduced at the *2005 ASMS Meeting on Mass Spectrometry and Allied Topics* held in San Antonio, Texas, June 5–9.

The Orbitrap has its origins in the device developed in 1922 by K. H. Kingdon at the Research Laboratory of General Electric Company in Schenectady, New York (Kingdon 1923). The Kingdon trap used an electrostatic field for trapping ions. It did not employ either a magnetic or an RF (dynamic electric) field. A radial logarithmic potential is created between two end-cap electrodes by applying a DC voltage between an outer cylindrical electrode and an electrically isolated thin wire that acts as a central electrode.

In 1981, R. D. Knight modified the shape of the outer electrode to add an axial quadrupole term to the equation of ion motion described by the original Kingdon trap (Knight 1981). This quadrupole potential "confines ions in the axial directions allowing them to undergo harmonic oscillations in the *z*-direction" (Hu 2005). Knight split the outer electrode radially on the middle to allow ions to be injected into the trap. Both the axial and the radial ion-signal resonance were found to be much weaker than that observed in the 3-D QIT. No mass spectra were reported for the "Knight-style" Kingdon trap.

R. Graham Cooks states that the "Orbitrap is a 'new mass analyzer'; however, for didactic purposes it is useful to consider the Orbitrap as a modified 'Knight-style' Kingdon trap with specially shaped inner and outer electrodes. Similarly, it can also be considered as a modified form of quadrupole ion trap although the Orbitrap uses static electrostatic fields, [whereas] the quadrupole ion trap uses a dynamic electric field typically oscillating at ~1 MHz. The Orbitrap's axially symmetric electrodes create a combined 'quadrologarithmic' electrostatic potential" (Hu 2005). Like the FTICR mass spectrometer, ion detection in the Orbitrap is carried out via broadband-image-current detection of a time-domain signal that is converted to a spectrum of mass-to-charge ratios and ion abundances using fast Fourier transform algorithms developed by Alan Marshall (Senko 1996).

The Orbitrap has *m/z* range and resolving power capabilities that make it a powerful analyzer. The Cooks' *J. Mass Spectrom.* paper reports a resolving power of 150,000; an *m/z* range of 6000; an *m/z* accuracy at high mass of 2–5 ppm; and a dynamic range of 10^3. The radial, angular, and axial frequencies are all *m/z* dependent; however, the axial frequency is completely independent of the energy and spatial spread of the ions in the trap. Because of the energy independence of the axial frequency of ions, the ability to separate ions and accurately determine the *m/z* values is possible. The high resolving power is due to the ability to accurately define the trapping field, a larger trapping volume compared to the 3-D QIT or the FTICR-MS, and an increase in space-charge capacity.

Ions are accumulated, cooled, and pulsed into the trap. Once injected, these ion bunches start coherent axial oscillations without the need for any additional excitation. All ions have exactly the same amplitude, but ion packets of different mass-to-charge ratios will execute their axial oscillations at their respective frequencies. The detection of an ion-image current due to motion along the z axis is only possible as long as the ion packets retain their spatial coherence in the axial direction. The outer electrode is split in half, allowing the ion-image current due to axial motion to be collected. The operating pressure in the Orbitrap is about 10^{-8} Pa. Even at this pressure, the ion packet can suffer a loss of phase coherence (undergo dephasing) through collision with background molecules that results in a decrease in signal intensity. The transient is Fourier transformed to create the mass spectrum.

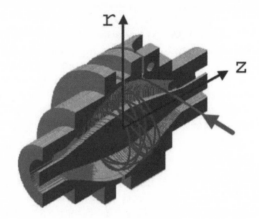

Figure 19. Cutaway of the Orbitrap mass analyzer. Ions are injected at the point indicated on the outer electrode. Ions are injected with a velocity perpendicular to the z axis. Ion injection at this point on the z potential is analogous to pulling back a pendulum bob and then releasing it to oscillate. (Hu, Q; Noll, RJ; Li, H; Makarov, A; Hardmanc, M; Cooks, RG "The Orbitrap: A New Mass Spectrometer" *J. Mass Spectrom.* **2005**, *40*, 430–433)

time-of-flight (**TOF**) mass spectrometer: This type of instrument uses no external force to separate ions of different *m/z* values. Ions are accelerated into a flight tube at a few hundred to several thousand volts (as high as 30 kV). The ions have differing velocities based on their mass. Lighter ions reach the end of the flight tube and are detected before the heavier ones. The equation that describes the ion separation in a TOF mass spectrometer is

$$m/z = 2Vt^2/D^2$$

where *m* is the mass of the ion, *V* is the acceleration voltage of the ion, *D* is the length of the flight tube, and *t* is the time from which the ion is accelerated until it reaches the detector at the end of the flight tube. The inherent resolving power of the TOF-MS is not very great; however, it can be enhanced with techniques of orthogonal ion injection and the use of reflective ion mirrors (see **Terms Associated with Time-of-Flight Mass Spectrometers**). The TOF-MS was developed in the late 1950s, was very popular in the 1960s and 1970s, fell into almost total disuse in the 1980s (except in Russia), and saw a resurgence in the 1990s because of MALDI and rapid data acquisition techniques.

The modern TOF-MS is used in GC/MS and LC/MS, as well as MALDI instruments, and as the second *m/z* analyzer in hybrid instruments where the transmission quadrupole, the quadrupole ion trap, or the double-focusing mass spectrometer is used as the first analyzer. In GC/MS, high resolving power (~10K) and high-speed data acquisition (500 Hz over an *m/z* range of 500) are commercially available.

Because ion separation is accomplished without the use of magnetic or electric fields using only the velocity of the ions, there is no theoretical limit to the *m/z* range for the conventional TOF called the **linear TOF** as opposed to operation with a reflectron referred to as the *re***TOF** mass spectrometer.

Note on measured accurate mass versus resolution (**resolving power**): An accurate *m/z* value can be measured regardless of the instrument's resolving power. Accurate *m/z* values can be determined using transmission quadrupoles or quadrupole ion traps where the resolution is no better than *m/z* 0.1–0.3. Accurate mass measurements are compared against calculated exact mass values for various combinations of elements to determine an elemental composition. This is why the average mass value is used when the resolution of the data is not sufficient to identify a monoisotopic *m/z* value. When a measured exact mass value does not compare favorably with various calculated exact mass values, it may be possible to show that the measured accurate mass peak is actually two mass spectral peaks that have become merged. This can be done by reducing the differences between measured *m/z* values (operation at a higher resolving power).

ion guide – a device that focuses ions into an *m/z* analyzer or collision cell. Ion guides are often an arrangement of round rods. In many cases, the ion guide is an **octupole,** which is four electric dipoles or four magnetic dipoles arranged to give a system with a zero net-dipole moment and a zero net-quadrupole moment. For a number of reasons, octupoles have also been called **octopoles** and **octapoles**, both of which are **incorrect** when referring to the device used as an ion guide. **Hexapoles** have also been used as ion guides. Some ion guides use a process called **collisional cooling** to help transmit more ions into the analyzer. More recently, a lens stack, such as that which makes up the **T-wave** technology used by Waters in their collision cells and ion transfer systems between an API interface and the first *m/z* analyzer, has been used as an ion guide.

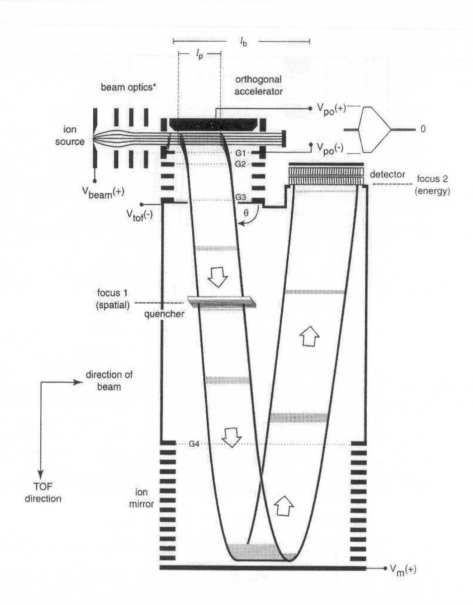

Figure 20. Schematic representation of a reflectron TOF mass spectrometer.
(Guilhaus, M; Selby, D; Mlynski, V "Orthogonal Acceleration TOF-MS" *Mass Spectrom. Rev.* **2000**, *19*, 65–107)

hybrid tandem mass spectrometer: a tandem-in-space MS/MS instrument that uses two different types of *m/z* analyzers (e.g., Q-TOF (a transmission quadrupole used as MS_1 followed by a collision cell followed by a time-of-flight analyzer used as MS_2) or a QIT-TOF (a 3-D QIT used as a device to isolate a precursor ion and have it undergo collision activation with the storage of the resulting product ions followed by a time-of-flight instrument used as MS_2). In addition to these two currently commercially available instruments, there is an instrument that uses a linear quadrupole ion trap as MS_1 and a collision cell followed by an FTICR-MS used as MS_2 and a linear quadrupole ion trap used as MS_1 and a collision cell followed by an Orbitrap used as MS_2. No longer commercially available, there have been hybrid instruments manufactured that consisted of a double-focusing instrument as MS_1 and a TOF as MS_2 and a system using a double-focusing instrument as MS_1 and a transmission quadrupole as MS_2. Although it may stretch the term "hybrid instrument", there were several systems manufactured by at least three companies that consisted of a reverse-geometry double-focusing instrument as MS_1 and a forward-geometry double-focusing instrument as MS_2. It should be pointed out that it is possible to perform all types of MS/MS (common-precusor-ion analysis, common-product-ion analysis, and common-neutral-loss analysis) in a double-focusing instrument using the linked-scan technique.

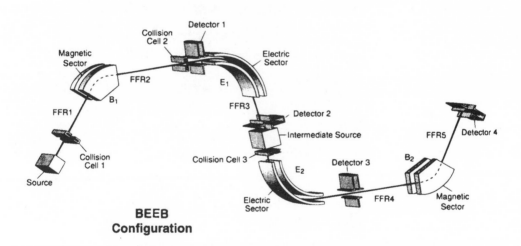

**BEEB
Configuration**

Figure 21. Schematic representation of a hybrid tandem-in-space mass spectrometer with a forward-geometry double-focusing instrument as MS₁ and a reverse-geometry double-focusing instrument as MS₂. (de Hoffmann, E "Tandem Mass Spectrometry: A Primer" *J. Mass Spectrom.* **1996**, *31*, 129–137)

TERMS ASSOCIATED WITH TRANSMISSION QUADRUPOLE AND QUADRUPOLE ION TRAP MASS SPECTROMETERS

AGC: **automatic gain control** (a.k.a. **ICC: ion current control**) is a process that controls the time that ions are formed in or accumulated in the trap. This process is to avoid space charge from an overload of ions in the trap. AGC is the term first used by Finnigan Corporation (now Thermo Electron) in describing the technique by which ion abundances are determined before ions are formed in or accumulated in the trap. Agilent Technologies and Bruker Daltonics use the ICC term.

axial modulation: This is another term coined by Finnigan Corporation referring to the application of an RF voltage to the end caps of a three-dimensional quadrupole ion trap that is half the frequency of that applied to the ring electrode. This reduces the effect of space charge and increases the number of ions that can be stored in the trap while still maintaining sufficient resolution to obtain an interpretable mass spectrum.

buffer gas: This is also called the **bath gas** or the **cooling gas**. This gas is He in most cases although Shimadzu uses Ar in a hybrid QIT-TOF LC/MS instrument. In the 3-D QIT, the buffer gas is used to cool the ions so that they will maintain individual orbitals for ions of differing *m/z* values. The pressure of this gas is critical and is maintained at approximately 10^{-1} Pa. In some GC/MS systems, the buffer gas is provided as the eluate from the GC column.

internal ionization: This is the formation of ions in the trap by electron or chemical ionization. This is only done with QITs used as a GC-MS.

electrodes: A QIT is constructed of three different electrodes: two **end-cap electrodes** and a **ring electrode**. These three electrodes, electrically isolated from one another, are used to construct the three-dimensional quadrupole field.

external ionization: This is the formation of ions in an ion source outside the trap followed by injection of the ions into the trap. This technique is used for both the GC-MS and the LC-MS. The LC-MS can only have external ionization.

hybrid CI: CI reagent ions are formed in a conventional ion source on a QIT GC-MS. These reagent ions are injected into the trap where they undergo ion/molecule reactions with analytes that are introduced in the same way as they are in the internal ionization systems. The reagent ions can be *m/z* value selected to give more specificity.

Lissajous trajectory: This is the shape of the orbital of an ion of a given *m/z* value in the 3-D QIT. In the common case, a Lissajous figure is the result of two periodic motions of the same frequency at right angles to one another. The following figure is for showing the motion of an aluminum microparticle trapped in a three-dimensional plane. This is a representation of the orbital for an ion of a single *m/z* value inside the 3-D QIT. (Todd, JKG; Lawson, G; Bonner, RF "Quadrupole Ion Traps" in *Quadrupole Mass Spectrometry and Its Applications*; Dawson, PH, Ed.; Elsevier: Amsterdam, 1976, pp 183–224).

An ion's Lissajous orbital

ass-selective instability: This is the mode of operation used by all 3-D QIT mass spectrometers. Developed at Finnigan Corporation by George C. Stafford and his research team in the early 1980s (Kelly 1983), this mode of operation involves the storage of ions of all *m/z* values from a lower limit defined by the magnitude of a fixed-frequency RF voltage applied to the ring electrode. By ramping the RF voltage magnitude, ions of consecutive *m/z* values become unstable and exceed the boundaries of the trapping field. The ions pass through the opening in one or both of the end caps and strike a conventional ion detector (electron multiplier or photomultiplier device). This technique provides the speed, resolution, and simplicity that has allowed the QIT to become such an important device in LC/MS and GC/MS. In addition to ramping the RF voltage on the ring electrode, mass-selective instability can also be accomplished by changing the DC voltage and the RF frequency individually or in combination with one another and the RF amplitude.

Mathieu diagrams: These are diagrams showing the solutions to the Mathieu equations that correspond to ion trajectories (stable or otherwise) and are displayed as a function of the mass and charge of the ions as well as the instrument's operating voltages. These diagrams and equations relate to ions that are stored by or transmitted through quadrupole fields.

ple-resonance ejection: Through the use of a hexapole resonance, a paramagnetic resonance due to a supplemental quadrupole field, and a supplemental dipole field on an ion cloud displaced from the geometric center of the ion trap, ions will absorb nonlinearly from all three fields simultaneously increasing the ion's axial motion resulting in rapid ejection with improved resolution.

waveform: This is a term used to describe the different voltages applied to the electrodes of three-dimensional quadrupole ion traps. These waveforms are used in the manipulation of stored ions and in selecting ions of specific *m/z* values for storage to the exclusion of ions of all other *m/z* values.

TERMS ASSOCIATED WITH DOUBLE-FOCUSING MASS SPECTROMETERS

decade: This term is used in conjunction with magnetic-sector and double-focusing mass spectrometers to describe an order-of-magnitude change in the *m/z* data acquisition range. Magnetic-sector mass spectrometers that obtain a spectrum using a fixed-field magnet and vary the accelerating voltage (no longer commercially produced) should only reduce the initial voltage by an order of magnitude. Lower values will result in decreased ion energy that will cause decreased sensitivity at the high end of the spectrum. A change in the magnetic field strength allows a different *m/z* decade range to be acquired for the same decade reduction in voltage. If an acquisition range of *m/z* 30–550 is required, then two separate scans have to be carried out using two different magnetic field strengths—one from *m/z* 30–300 and the other over the range of *m/z* 60–600. Although the first would require the full-decade scan time, the second would only require the fraction that is needed to acquire data from *m/z* 301–550; however, some overlap should be used to assure that a complete spectrum is obtained.

In modern instruments, the magnet can be scanned linearly (incrementing a DAC; only done by JEOL) or exponentially:

$$M = M(0)e^{kt}$$

where *M*(0) is the starting *m/z* value, *k* is a constant, and *M* is the *m/z* value at the time *t*. More than a decade can be scanned, but the scan-time equation uses the time to scan a decade (20–200, 60–600, etc.) to define the scan speed for an exponential scan, which has the advantage that all peaks have a constant width in time regardless of *m/z* value. An exponential magnet downscan is also easy to implement in an analog fashion by discharging a capacitor. Linear magnet scans emphasize higher mass peaks. Linear scans can start at *m/z* 0, but *m/z* 0 can never be obtained with an exponential scan.

forward geometry: This term describes a double-focusing mass spectrometer in which the electric sector follows the ion source and precedes the magnetic sector. This instrument has an **EB** configuration.

reverse geometry: This term describes a double-focusing mass spectrometer in which the magnetic sector follows the ion source and precedes the electric sector. The electric sector is the last field that is experienced by the ions prior to their entering the detector or a collision cell. This instrument has a **BE** configuration.

sector field: This **incorrect** term is a redundant and confusing neologism that has been used to describe a mass spectrometer that consists of a magnetic and electric sector (a double-focusing instrument). What is a *sector field*? Both the magnetic portion of the double-focusing mass spectrometer used for velocity focusing and the electric portion used for energy focusing are considered sectors; and both provide fields, magnetic and electric, respectively. Both are referred to as sectors. The use of the term *sector field* can only lead to confusion when trying to describe a mass spectrometer. If the term describes an instrument where there is a sector that is also a field, then it should be hyphenated (i.e., a sector-field instrument). The common use of the term *sector field* is applied to double-focusing instruments. However, would not a *single-focusing* mass spectrometer be a *sector-field* instrument? The term *double focusing* (which has been in use for over 65 years) used to describe instruments that use separate sectors that provide magnetic and electric fields does not need to be replaced or have a synonym.

field-free region: This term (abbreviated **FFR**) is not necessarily limited to sector-based mass spectrometers. It is any region of the mass spectrometer where the ions are not subjected to a field (electric, magnetic, quadrupole, etc.). Avoid using arcane terms such as "first FFR" or "second FFR". Use "…the field-free region traversed by the ion beam before entering the electric sector" or "…the field-free region between the electric and magnetic sectors."

MIKES and **IKES**: These abbreviations are for **mass-analyzed ion kinetic energy spectrometry** and **ion kinetic energy spectrometry**, respectively. These techniques are used to obtain a product-ion mass spectrum of a metastable transition or product ion produced from a precursor ion in a collisionally activated dissociation. MIKES can only be accomplished in a reverse-geometry double-focusing mass spectrometer (Watson 1997).

linked scan: This is a term that refers to the dependence of the magnitude of the electric sector (deals with energy dispersion of ions) on that of the magnetic sector or accelerating voltage out of the ion source (deals with the momentum dispersion of ions) to one another in order to obtain data on metastable transition in a double-focusing instrument. The linked scan is used to carry out MS/MS experiments using a double-focusing mass spectrometer by placing collision cells in various field-free regions of the instrument.

There are basically five types of linked scan: **linked scan at constant B/E** and **at constant E^2/V**, which results in a common-precursor-ion analysis; **linked scan at constant B^2E** and **at constant B^2/E**, which results in a common-product-ion analysis; and **linked scan at constant $B[1 - (E/E_0)]^{1/2}/E$**, which results in a common-neutral-loss analysis. In each of these techniques, **B** is the magnetic field strength, **E** is the electrostatic field strength, and **V** is the accelerating voltage. The relation of B and E is a function of the *m/z* value of the common precursor or common product ion.

Common-precursor-ion analyses using the linked scan at constant B/E result in sharper mass spectral peaks (better resolution) than linked scan at constant E^2/V. Common-product-ion analyses using linked scan at constant B^2E result in sharper mass spectral peak shapes than linked scan at constant B^2/E. Due to design limitation of the hardware, there is a lower limit to the *m/z* values of product ions that can be detected using this technique. This so-called low-mass cutoff is something that the link scan has in common with MS/MS in the quadrupole mass spectrometer. Both the constant E^2/V and the constant B^2E scans can be carried out in either a forward-geometry or reverse-geometry instrument. The collision cell is located in the field-free region traversed by the ion beam before entering the electric sector. This results in product ions having the same velocities as the precursor ions but different momenta and different kinetic energies. The terms *E^2/V scan* and *B^2E scan* should not be used because they imply that B and E and E and V vary with respect to one another during the scan.

The linked scan at constant B^2E produces sharper peaks than produced with constant B/E. However, the linked scan at constant B^2E requires a reverse geometry with a collision cell between the electric and magnetic sectors (the second FFR). The linked scan at constant B/E can be carried out in either a forward-geometry or reverse-geometry instrument with the collision cell in the field-free region traversed by the ion beam before entering the electric sector. The linked scan at constant B^2E is accomplished by scanning the magnetic sector from low to high field at the same time the electric sector is scanned down in such away as to maintain B^2E as a constant.

The linked scan at constant B^2/E produces a common-product-ion analysis from the the field-free region traversed by the ion beam before entering the electric sector with poorly resolved mass spectral peaks. The resolution is sufficient to identify a homologous series of compounds.

main beam ions: This term refers to the ions that are in the **main ion beam**, which is a beam of ions (of multiple *m/z* values) that has a kinetic energy equal to the full accelerating energy.

TERMS ASSOCIATED WITH TIME-OF-FLIGHT MASS SPECTROMETERS

coaxial reflectron: This term refers to a TOF instrument in which the ion source, reflectron, and detector are on the same axis (in line with one another). Ions travel in a line from the ion source to the reflectron, back in the direction from which they entered the reflectron through the ion source, and into the detector, which is in front of the ion source.

curved-field reflectron: This type of device allows for the energy focusing of ions over a broad *m/z* range to produce a complete mass spectrum from a single laser shot that induced postsource decay of a large molecular mass analyte (i.e., proteins).

draw-out pulse: (a.k.a. *push-out pulse*) This term is used in reference to the voltage used to remove the ions from the ion source in a TOF mass spectrometer. This process of ion removal is accomplished by the pulsing of the ion-extraction field.

delayed extraction: (originally known as **time-lag focusing**) The delayed extraction technique is the time delay between the ionization pulse and the draw-out pulse. This technique is used to improve mass resolving power.

ion gate pulse: This technique is the application of an electrical pulse between either the ion source and the *flight tube* (a.k.a. *field-free region* or *drift tube region*) of the TOF mass spectrometer or the flight tube and detector to allow ions of only a narrow *m/z* range to pass.

gridless reflectron: Because there is scattering of ions after passing through grids (an actual wire mesh) that make up the ion mirror in a TOF-MS (which can result in ion losses ~10%), the gridless reflectron was developed. The gridless reflectron uses a series of rings with varying potential applied to create the mirror rather than the grids. Tight focusing of the ion beam close to the reflectron axis is required because divergence of ion trajectories for off-axis ions can result.

orthogonal extraction: Ions are pushed into the extraction region. A pulse is applied at a 90° angle to the direction of the ion flow to push the ions into the flight tube of the TOF-MS. A common abbreviation for a TOF mass spectrometer employing orthogonal ion injection is *oa*TOF.

postsource decay: Postsource decay (PSD) is the fragmentation of ions (once they have been fully accelerated) in the drift tube prior to entering the reflectron, if one is present. Although postsource decay is a term often used in conjunction with time-of-flight mass spectrometry, the term does not depend on the style or pressure of a certain analyzer—it is an ion property.

reflectron: The reflectron (referred to as an ion mirror in its simplest form) is used to change the direction in which the ions are traveling (reflecting field) and to energy-focus ions (retarding field) for improved resolution. A single-stage reflectron has a single-retarding reflecting field, whereas a dual-stage reflectron has two linear-retarding voltage (constant field) regions that are separated by an additional grid. A dual-stage reflectron allows for smaller flight tube lengths compared to single-stage reflectrons. A common abbreviation for TOF instruments that have the reflectron is *re*TOF.

transient: a continuum of ion current consisting of the sequential arrival at the detector of individual packets, representing ions of progressively increasing *m/z* value (i.e., it is the complete mass spectrum derived from a single extraction pulse of the ion source in a TOF mass spectrometer).

The following terms are associated with MALDI MS. Although any type of *m/z* analyzer can be used with MALDI, the TOF-MS is most often employed.

fluence: This term is associated with laser-induced ionization such as in MALDI. This term is borrowed from physics, but it has a somewhat different definition in mass spectrometry. In mass spectrometry, fluence relates to the energy of the laser, the size of the impact area, the cross-sectional area of the sample, and the time of the laser pulse. In physics, energy fluence (Ψ) is expressed as a function of time (i.e., the energy fluence rate or energy flux density).

ablation: As used with reference to mass spectrometry, this term refers to the process of removing particles from the surface with a laser. In mass spectrometry, the removed (laser-ablated) molecules are ionized by charge transfer (one theory for what takes place in MALDI), photoionization, or through energy from a second laser. In MALDI, it has been theorized that the analyte molecules entrained in the matrix plume, which is undergoing a supersonic expansion, are ionized by the matrix molecules that have been ionized while still in the solid phase in the matrix.

Ablation is one of a series of words taken by mass spectrometry to describe a phenomenon that happens in the technique that has no single describing term. As defined in the dictionary, ablation means: (1) surgical excision or amputation of a body part or tissue; (2) the erosive processes by which a glacier is reduced; (3) in aerospace, it is the dissipation of heat generated by atmospheric friction, especially in the atmospheric reentry of a spacecraft or missile by means of a melting heat shield. From Latin *ablātus*, past participle of *auferre*, to carry.

raster: Often used to describe the movement of a laser over a target or an ion beam on a detector. This term describes the movement one increment at a time where each new position is defined as the movement from one set of *x, y* coordinates to another with the difference between each set always being the same.

USE OF ABBREVIATIONS

Use of abbreviations in books, reports, and articles is a necessity in order to have an easy-to-read document and to have a document that is not excessive in length. Some reports on mass spectrometry terminology list a series of abbreviations that can be used without definition. All abbreviations should be defined before their use. In lengthy documents, an abbreviation should be defined multiple times in order to produce clarity. Nothing is more frustrating than to follow an index to a page in the middle of a 500-page book that contains several esoteric abbreviations that were defined in several isolated places in previous chapters. In such cases, it may be advisable to include a table of abbreviations. In a recent book on mass spectrometry, such a table had more than 100 items, which allowed for a reasonable size book and facilitated its use as a reference (de Hoffmann 1996). The following is a guide to some abbreviation suggestions.

A great deal of care must be taken in the use of abbreviations to describe instruments such as a **GC-MS** and analytical techniques such as **GC**, **MS**, and **GC/MS**. This care is especially true with respect to "hyphenated instruments" and with techniques that use hyphenated instruments. **GC** or **LC** is almost always used to refer to an instrument—a **gas chromatograph** is a **GC** and a **liquid chromatograph** is an **LC**. However, **GC** can also be used to refer to the technique of **gas chromatography**; **LC** can also be used to refer to the technique of **liquid chromatography**.

Sometimes, **MS** is used to refer to an instrument—a **mass spectrometer** (e.g., the time-of-flight MS can have good mass accuracy); other times, it is used to refer to a piece of data—a **mass spectrum** (e.g., the MS of cocaine is quite distinctive); or it can also be used to refer to the technique of **mass spectrometry** (e.g., MS is one of the most used analytical techniques). Care must be taken in the use of these abbreviations to clarify what is meant. It would be considered bad form to use one of these abbreviations to mean two or more different things in the same document. Because the use of the words *mass spectrometer*, *mass spectrum*, and *mass spectrometry* make for long documents and tedious reading, one of the three is often abbreviated. The word *spectrum* can be substituted for mass spectrum. The use of the MS abbreviation for mass spectrometer and mass spectrometry in the same document can be understandable by those with English as a first language; but for those with some other first language, this convention can be confusing. Abbreviations should be defined the first time they are used, and then used as often as required for clarity when it is not possible to include a table of abbreviations.

Some references have been made to "hyphenated techniques". **Hyphenated techniques** should mean techniques that use **hyphenated instruments**. The slash (/) is used to separate a combined technique such as gas chromatography/mass spectrometry (GC/MS) because the result of the chromatography is being analyzed by the mass spectrometer. Although the *Current IUPAC Recommendations* specifically states that GC/MS and LC/MS can be used without definition for either the combined techniques or the combined instruments, this practice can create confusion and should be avoided. All abbreviations should be defined.

The term "hyphenated methods" was first coined by Tomas Hirschfeld (Hirschfeld 1980). In this presentation, Hirschfeld talked about the hyphen being the interface between two instruments, one being ideally suited for quantitation and the other being ideally suited for qualification. He also noted the necessity for compromise needed to develop a method using these "hyphenated instruments". As an example, it is important to remember that there is as much difference between LC/MS (the technique) and either liquid chromatography or mass spectrometry as there is between the individual techniques of liquid chromatography and mass spectrometry; however, this unique analytical technique of LC/MS is accomplished with an instrument that is both a liquid chromatograph and a mass spectrometer. This instrument is hyphenated because of the interface between the two. The technique of LC/MS is *not* both a liquid chromatographic and a mass spectrometric technique at the same time. Special considerations that do not pertain to liquid chromatography or mass spectrometry must be made in the development of an LC/MS method that uses an LC-MS.

GC/MS is used as the abbreviation for the technique of gas chromatography/mass spectrometry. **GC-MS** is used as the abbreviation for a gas chromatograph-mass spectrometer because the instrument combines a mass spectrometer with a gas chromatograph. Therefore, **LC/MS** refers to liquid chromatography/mass spectrometry; **LC-MS** refers to a liquid chromatograph-mass spectrometer. **CE/MS** refers to capillary electrophoresis/mass spectrometry; **CE-MS** refers to a capillary electrophoresis-mass spectrometer.

A single analytical technique is abbreviated without a hyphen or a slash (/). Therefore, when using **MS** as an abbreviation for mass spectrometry, the correct abbreviation for time-of-flight mass spectrometry would be **TOFMS**; however, when using **MS** as the abbreviation for a mass spectrometer, the correct abbreviation for a time-of-flight mass spectrometer would be **TOF-MS** (the two are separated by a hyphen).

When the ionization method is used in the description of the mass spectrometer, the ionization method and the instrument type are separated by a space (i.e., **EI TOF-MS** for an electron ionization time-of-flight mass spectrometer). A space following the ionization type would also be correct for inclusion with the ion-separation technique (i.e., **EI TOFMS** for electron ionization time-of-flight mass spectrometry).

The above stated use of the hyphen and slash is supported by the Information to Authors section in recent issues of *Rapid Communications in Mass Spectrometry.* Other mass spectrometry journals do not specifically address this issue in their author-instruction sections.

The technique of MS/MS involves more than a single *m/z* analyzer, except in the case of quadrupole ion trap (QIT) and Fourier transform ion cyclotron resonance (FTICR) mass spectrometers. Multiple *m/z* analyzer instruments are sometimes described with an abbreviated notation that uses the single-letter symbols for the instrument type described above. The lower case **q** is a single-letter symbol used to describe the location of an RF-only quadrupole collision cell. Some examples are **QqTOF** for a transmission quadrupole *m/z* analyzer, followed by a quadrupole collision cell, followed by a time-of-flight *m/z* analyzer; and a **QqQ** as the abbreviation for a triple-quadrupole mass spectrometer.

A good example of problems created by the use of abbreviations is **TMS**. In recent literature, TMS has been used as an abbreviation for tandem mass spectrometry. TMS has long been considered by gas chromatographers and mass spectrometrists the accepted abbreviation for **trimethylsilyl** derivatives, as in TMS derivatives of alcohols. This, as with all abbreviations, should not be used without definition; and TMS should *not* be used as an abbreviation of trimethylsilyl and tandem mass spectrometry in the same article. It is best to avoid the use of TMS as an abbreviation for tandem mass spectrometry.

When to use alpha, α, alpha-, or α-, (etc.)
An important nomenclature convention is the use of spelled-out Greek words as opposed to the Greek letters (e.g., alpha versus α). When the nomenclature **alpha**, **beta**, and **sigma cleavage** was first proposed (Budzikiewicz 1964), the authors incorrectly elected to use chemical-name conventions. Just as β-naphthol was written with a hyphen to indicate that the functionality is at the beta position of the molecule, α-cleavage was written with a hyphen. This style placed the index reference for α-cleavage under **C**. Subsequent authors (McLafferty 1993; Watson 1997) refer to the breaking of the alpha bond as alpha cleavage in the index and place the reference under **A**, which is more correct as an ease-of-use feature. In the case of alpha cleavage, alpha is an adjective that modifies cleavage, which is a noun; therefore, alpha cleavage is not a hyphenated term. These latter authors, who indexed alpha cleavage, use the term "α-cleavage" **incorrectly** in the text of their books. If the symbol is used, there is no need for a hyphen because the symbol is used as a shorthand form of the adjective (alpha) that modifies the noun (cleavage). The α does not specify the location of a functionality in a molecule or an ion. The **correct** term is α cleavage. However, when α cleavage is used to describe an event (e.g., an α-cleavage reaction), then a hyphen is used because α-cleavage is an adjective that describes reaction, which is a noun.

From the early mass spectrometry literature through the present, the words and the Greek letters were used for alpha, beta, and sigma. The first edition of *The ASC Style Guide* (Dodd 1986) states: "Greek letters, not the spelled-out forms, are used in chemical and drug names."

Unfortunately, the examples that are given use the hyphen with the Greek letter and omit the hyphen when the spelled-out name is used (e.g., β-naphthol versus beta naphthol). In this example, a hyphen should have been used regardless of the chosen form because beta is a prefix, *not* an adjective.

According to the second edition of *The ASC Style Guide* (Dodd 1997), the Greek letter is the preferred style in ALL cases: "Use Greek letters, *not* the spelled-out words, for chemical and physical terms."

The examples given in this case are γ radiation versus gamma radiation and β particle versus beta particle. This particular style change was probably a result of the desire for less print in articles but presents a dilemma for mass spectrometry. For the sake of clarity, the spelled-out word should be used where needed. If a Greek letter (e.g., α, β, or σ) is used to describe a cleavage, then no hyphen is necessary when the letter is meant as a shorthand notation for the word used as an adjective that modifies cleavage.

Remember, a γ-hydrogen (the γ is a prefix) shift-induced cleavage requires the use of a hyphen, whereas no hyphen is required to describe the γ hydrogen (the γ is a shorthand notation for gamma, an adjective) that has shifted to bring about the cleavage.

COMPONENTS OF A MEASUREMENT

There are four components to a measurement: (1) signal strength, (2) baseline offset (offset from zero), (3) background (signal strength due to background), and (4) noise (reproducibility of the signal from one sample to the next).

signal strength a measured response of the system to a precise amount of sample.

baseline offset the difference between zero of the system and the lowest level that a signal can be observed when no sample is present.

background the level and variation of the signal when ionization is on, but no sample is present.

noise the variation of signal strength when a sample is present from one measurement to the next.

At the lower levels of detection, the noise and the background will be close enough to the same value to be considered equal; therefore, the detection limit of a system can be defined as the ratio of the signal strength produced by the sample to the variation from one measurement to the next (**signal-to-noise**) or the ratio of the signal strength produced by the sample to that of the background for a measurement (**signal-to-background**). If signal-to-background is used as the measurement of sensitivity, then the background signal must be measured in the same sampling interval as the sample signal and should be measured as near as possible to the sample signal. Whether sensitivity is determined as an S/N or S/B, the value for the zero offset must be taken into account.

The **limit of detection** is the minimum amount of sample that can produce a signal that allows the determination of the presence of the sample and is expressed as an S/N or S/B value relative to an amount (e.g., 5 pg of hexachlorobenzene will produce a mass chromatographic peak that has an S/N value greater than 10). Detection limits usually assume that only the specified analyte is producing the signal. In some cases, a detection limit may specify a criteria for confirming the presence of the analyte (e.g., in the above example, it may be stated that the sample must produce a library searchable spectrum).

The **limit of quantitation** is the smallest amount that can be measured to within a specified precision.

The precision of a numerical result is the degree of agreement between the result and other values obtained under the same conditions. The percent precision is 100 divided by the smallest numeric value that can be measured under a given set of conditions (e.g., a 50% precision will allow a value as small as 2 to be measured, a 20% precision will allow a number of no less than 5 to be measured, and a 5% precision will allow the measurement of a number of no less than 20).

The limit of quantitation becomes the smallest number that can be measured times the noise. If the number of units of noise in a measurement is 5 and a 5% precision is required for the limit of quantitation, then the smallest measurement allowed is 100 units. It should be remembered that the level of background and the level of noise are the same at the lowest signal level.

FORMULAS AND EQUATIONS

If a formula or an equation contains mathematical operators (+, −, =, ×) that represent an arithmetic function where the + and − does not relate to the sign of a number (or symbol) or the result of the product of a set of numbers (or elements), then the operator should be preceded and followed by a space. No spaces precede or follow the slash (/) when used as a mathematical operator. Parentheses and brackets should be presented without preceding or trailing spaces, except when using the arithmetic operators.

Examples:

$E = mc^2$

NOT $E=mc^2$ or $E = m\ c^2$

$A + B = C$

NOT $A+B=C$ or $A+B = C$

$(a/b)(c/d)$

NOT $(a\ /\ b)(c\ /\ d)$

$1 + [2A − B] = [C + D]$

NOT $1+[2A−B]=[C+D]$ or $1 + [\ 2A − B\ ] = [\ C + D\ ]$

$PV = nRT$

NOT $PV=nRT$

To be consistent with IUPAC recommendations, the symbols in equations should be in italics, but the operators should not be in italics.

In the presentation of chemical equations, **reaction symbols** are used. These four sets of arrows have specific meanings in the presentation of a chemical reaction. (1) Items on the left of the arrow are the reactants and starting materials, and those on the right are products. (2) When a reaction is reversible, to show that the reactants' concentrations can be in equilibrium with one another, half arrows pointing in opposite directions are used. (3) If at equilibrium and either the reactants or the products dominate the mixture, then a larger arrow is used to indicate which is the dominant. (4) In the case where two or more structures are used to indicate **resonant** forms, the possible structures are separated by a double-headed arrow. Of course, none of these resonant forms is a correct structure because instead of the bonding pair of electrons being localized between two atoms, the bonding pair of electrons is spread over several atoms.

| 1 | 2 | 3 | 4 |

TYPES OF ELEMENTS AND ELECTRONS

For the purpose of describing the various types of elements with respect to the isotopic ratios that are used in the determination of the elemental composition of an ion, the common elements found in organic mass spectrometry are referred to as:

X elements those elements that exist as only a single naturally occurring stable isotope (H, F, P, I). Because of the low natural abundance of deuterium compared to hydrogen, hydrogen is considered an X element.

X+1 elements those elements that exist as only two naturally occurring stable isotopes, and that are 1 integer u different (C and N).

X+2 elements those elements that exist as more than a single naturally occurring stable isotope, and two of the isotopes differ in integer value by 2 u (Cl, Br, S, Si, and O).

It should be noted that the letter **A** rather than **X** has been used to describe these different types of elements (McLafferty 1993). McLafferty's use of the symbol **A** is probably because **A** is the IUPAC symbol for mass number. However, because **X** is a universally accepted symbol for an unspecified number, and unspecified ions or peaks can occur with various values, the use of **X** is more descriptive and therefore is recommended.

An organic compound, molecule, ion, and/or radical has three types of electrons:

n- or **nonbonding electrons**, which are not associated with the attachment of one atom to another.

π or π**-bond electrons**, which are found in a chemical bond formed by the overlap of two adjacent *p* orbitals.

σ or σ**-bond electrons**, which are associated with bonds formed by the overlap of an *sp*, *sp^2*, or *sp^3* hybrid orbital, and another such hybrid orbital or an *s* orbital.

Historical Note: When Sir John Joseph Thomson discovered the electron in 1897 (Thomson 1897), he used an apparatus to deflect cathode rays in magnetic and electric fields in a high vacuum, thus avoiding the interference from the ionization of residual gases; this apparatus evolved into the mass spectrograph. He called these particles (with a mass 1/1845 of a hydrogen atom) that composed cathode rays *corpuscles* rather than *electrons*. George Francis FitzGerald (Irish physicist 1851–1901) proposed that the term "electron", coined by George Johnston Stoney (Irish physicist 1826–1911), FitzGerald's uncle (FitzGerald 1897), be used to describe this subatomic particle with a negative charge. Stoney first put forth the idea of atoms having a unit charge at a British Association for the Advancement of Science meeting in 1874 to describe electricity (Stoney 1881). In 1891, in a presentation before the Royal Society of Dublin, Stoney readdressed this issue when he said, "…these charges which it will be convenient to call the electron" (Stoney 1891). This was the first use of the term, which was in wide acceptance at the time of Thomson's discovery.

TERMS ASSOCIATED WITH COMPUTERIZED SPECTRAL MATCHING

Library Search

There are a number of data-system-specific terms that are used with various mass spectral library search programs. They can be reviewed in the documentation used with each manufacturer's data system. The following terms have appeared in the literature and are noteworthy.

fit: This term is specific to the INCOS library search system distributed by Finnigan and Varian. Peaks that are in the sample spectrum and not in the library spectrum are disregarded in evaluation of the match. This type of comparison has the advantage of being able to identify multiple compounds represented by a single spectrum.

match factor: This term is used to describe the quality of the match between the sample spectrum and the spectrum found by the NIST Mass Spectral Search Program for Windows (NIST MSS Program). It has the same meaning as *purity* in the INCOS search. It is based on the dot-product result of a point-by-point comparison of the two spectra.

purity: This INCOS-specific term (as distributed by Varian and Finnigan) is used instead of *similarity index*. It takes into account all peaks in the sample and library spectra to make a determination of how close the two spectra match. The purity is calculated by using mass-weighted intensities in conjunction with a linear-algebra dot product. For more information on this term, consult the data system manuals from Finnigan or Varian.

probability: This term has two different meanings. The first meaning is associated with the Probability Based Matching (PBM) algorithm as implemented by Agilent Technologies, Palisade Corporation, Teknivent, ThruPut Software, and others. This term is supposed to be a value that represents the probability that the "unknown is correctly identified as the reference". However, the HP G1034C MS ChemStation Software manual (ca. 1993) states, "Values less than 50 mean that substantial differences exist between the unknown and reference, and the match should be regarded with suspicion." This wording sounds like a comparison of the two spectra is being made, which would mean that the *probability* term is being used in the same way as the *purity* term of INCOS and the *similarity index* term of NIST (a probability that the two spectra are from the same compound based on a direct mass-and-intensity comparison).

The second definition of the term *probability*, as used in the NIST MSS Program, describes the likelihood that the mass spectrum of the unknown and the mass spectrum retrieved from the database are of the same compound. The probability is based on all the matches found during the search. In the case of a search of a submitted spectrum of *m*-xylene, there would be matching spectra of all three isomers of xylene found in the library. All three would have very high match factors when compared with the submitted spectrum, and would mean that the probability of any one of the three library spectra being of the same compound as that which produced the submitted spectrum would be divided among the three and would therefore be low. The NIST MSS Program is the only program that offers this type of comparison.

quality: This term is used in the Agilent Technologies implementation of the PBM mass spectral search algorithm to describe how closely the submitted spectrum and the library spectrum match. It is not clear from the HP documentation how the word *quality* used in this case differs from the *probability* value reported for a match.

(Rfit) reverse fit: *Rfit* is another INCOS-specific term. It means that the evaluation is based on the peaks in the submitted spectrum being in the library spectrum. The library spectrum can have additional peaks without affecting the value of the similarity.

reverse match factor: This term is unique to the NIST MSS Program. It is a *match factor* that is calculated by ignoring any peaks in the sample spectrum that are not in the library spectrum. These peaks are considered to be due to impurities. This term is similar in nature to the *fit* definition in the INCOS search and the *reverse search* value used in the NIST MSS Program for DOS.

reverse search: This term can also have two different meanings. The first meaning pertains to a sequential search of a data file of mass spectra to see if it contains a spectrum of a specific compound. This type of search is used in the analysis of GC/MS or LC/MS data for target compounds. The reverse search is the basis for many automatic search and quantitation programs. A retention-time window in the data file is searched to see if a spectrum is present that matches the spectrum of the target analyte.

The second definition of the term *reverse search*, as used in the NIST MSS Program, means the "hit list" is presented in decreasing order of the *reverse match factors* as opposed to decreasing order of *match factors*. This process treats the extraneous peaks in the submitted spectrum as impurities. The *reverse search* values in the NIST MSS Program are the same as the *fit* values used to describe the results of the INCOS library search provided by Finnigan and Varian.

similarity index (SI): This term has been used by a number of different mass spectral library search programs. The *similarity index* is a comparison of *m/z* values and relative intensities. SI was used with the NIST MSS Program for DOS but abandoned in the Windows version. There is another term that has also been used with a similar meaning: **dissimilarity index (DI)**. Both of these terms are used to describe how similar the library spectrum is to the submitted spectrum. Neither similarity index nor dissimilarity index is used with current commercial data systems.

Search Algorithms

There are primarily two search algorithms in use in most commercially available data systems: (1) the Probability Based Matching (PBM) algorithm (McLafferty 1974); and (2) the dot-product algorithm (a.k.a. INCOS [Sokolow 1978], which is associated with a Finnigan Corporation-specific data system). A number of enhancements and variations to the INCOS algorithm have been made and are distributed by the National Institute of Standards and Technology (NIST), Micromass, Finnigan, Varian, Agilent Technologies (on their GC/MS and LC/MS products), Perkin-Elmer (on their GC/MS product only), and others. The PBM algorithm is used by Waters (in their LC/MS system); and Teknivent, Shimadzu, and ThruPut in their GC/MS systems. The PBM algorithm is distributed by Agilent Technologies, along with the NIST version of the INCOS algorithm on their GC/MS products only. The variations in the implementation of the INCOS algorithm primarily have to do with the presearch and spectral condensations.

The presearch is a system by which possible candidates for a match to the unknown spectrum are selected from the mass spectral database. This presearch is carried out by using a subset of the unknown spectrum and the database spectra (one or more types of spectral condensations). Some presearch algorithms involve only a reduced sample spectrum (8 most intense peaks) and less condensed database spectra (16 most intense peaks). Other algorithms employ multiple criteria. The presearch is an important area of consideration. If the spectrum of the sample compound is excluded in the presearch, then the results are much like "throwing the baby out with the bath water". There have been other algorithms that existed in the formative years of mass spectral data systems. Some are included in the following brief descriptions.

PBM: The Probability Based Matching (PBM) algorithm uses a *weighting* and *reverse search*. The *weighting* is used to determine the importance of peaks based on *m/z* values and relative intensities. The weighting is used in the presearch. If the sample spectrum and the library spectrum of the same compound do not have the same base peak, then it is very likely that the library spectrum of the matching compound will be excluded. The probability that particular abundances will occur follows a log–normal distribution. The probability of most mass values also varies in a predictable manner. Because the larger molecular fragments tend to decompose to give smaller fragments, the probability of higher *m/z* values decreases by a factor of 2 approximately every 130 *m/z* units. This weighting system is used for indexing the "Important Peak Index of the Registry of Mass Spectral Data". The second feature of PBM, *reverse searching*, treats peaks in the submitted spectrum and not in the library spectrum as if they are from another compound. This process is valuable in identification of the spectrum of more than one compound. The PBM algorithm compares the submitted spectrum against the "Important Peak Index". Spectra found by this comparison are evaluated against the spectra in the entire database or a condensed version of it. Because of the way that the "Important Peak Index" is developed, the PBM search algorithm is limited in its performance when small databases of target compounds are considered.

INCOS: The INCOS library search system uses a presearch of matching the 8 most intense peaks in the submitted spectrum with a set of the 16 most intense peaks in the library spectrum. A search is performed against the 50 most chemically significant peaks in the spectra that are retrieved by the presearch. This search uses a weighting of the intensity so that peaks of higher *m/z* values are more significant than peaks of lower values. The comparison is done relative to adjusted abundances, which are the *square root* of the observed peak intensity *times* the *m/z* value. The results are reported for *purity*, *fit*, and *Rfit* (see above). Unfortunately, the condensation algorithm to produce the condensed database used by Finnigan and Varian is not very good with respect to which peaks are discarded. A condensation of a full-spectrum database (such as the Full.d file produced for the Agilent Technologies system) using **MassTransit**™ (Palisade Corporation) produces a far-superior condensed database because MassTransit uses the same *weighting* described with the PBM algorithm.

NIST: The library search system algorithm of the NIST MSS Program is similar to that of INCOS, except that the presearch and the main search are against a set of weighted intensities. The user has control over a number of different factors that can affect this search, such as whether or not *neutral loss* logic is used. This search program is offered by a number of different manufacturers in addition to their own routines.

N most int. Peaks: This algorithm is a comparison of a set of most intense peaks in the spectrum. This comparison is usually carried out against a set of eight peaks. This algorithm requires that the spectrum that corresponds to the submitted spectrum of the sample compound is in the mass spectral reference library. This algorithm has the advantage of being able to program easily, and the amount of disk storage space is limited.

N most sig. Peaks: This algorithm is a comparison of the most intense peaks after a weighting of intensities. This algorithm also requires that the spectrum that corresponds to the submitted spectrum of the sample compound is in the mass spectral reference library. This algorithm has the same advantages as the *N most int. Peaks* algorithm.

Biemann search: This algorithm is a comparison of the 2 most intense peaks within each window of 14 *m/z* units, starting the divisions at *m/z* = 6. This algorithm has the same limitations and advantages as the *N most int. Peaks* and *N most sig. Peaks* algorithms (Hertz 1971).

neutral loss: This algorithm requires that the matching spectra must have peaks that correspond to neutral loss peaks in the submitted spectrum. This type of search requires that the nominal mass of the compound of the submitted spectrum is known. The neutral loss search usually is used in combination with other algorithms and requires a database of neutral losses.

RTI & spectrum: The retention-time index (RTI) and spectrum-match algorithm is most often used with target-compound analyses. This type of search compares all the spectra that are in a given retention-time window and determines if there is a reconstructed chromatographic peak that contains spectra that have the mass spectral peaks found in the reference spectrum. This algorithm uses a *reverse search* because it does not consider peaks in the sample spectrum that are not in the library spectrum.

ION DETECTION

The ions that are *m/z* analyzed by a mass spectrometer are detected with several different devices:

dynode: (**high-energy** or **conversion**): A dynode is an element in an electron tube (electron multiplier: EM) whose primary function is to provide secondary emission of electrons. Secondary emission occurs when an electron that moves at sufficiently high velocity strikes the dynode. Positive and negative ions, as well as fast-moving neutrals, can also be used to produce an emission of secondary electrons. In a transmission quadrupole mass spectrometer (TQ-MS) and a quadrupole ion trap mass spectrometer (QIT-MS), higher-mass ions strike the detector with less velocity than lower-mass ions. This variation in velocity of ions can result in mass discrimination. A **high-energy dynode** is used to accelerate the ions so that the secondary emission caused by the impact of the ion on the electron multiplier's first secondary emission dynode is more equal regardless of its mass. Similar actions are taken on high-mass MALDI ions mass-resolved in a TOF-MS. The addition of a high-energy dynode has been used to produce the **postacceleration detector** (**PAD**).

A **conversion dynode** is used to convert negative ions to positive ions in the TQ-MS and QIT-MS. The first dynode of an EM has a negative bias to attract positive ions. Due to the low ion acceleration in a TQ-MS (10–15 V), the negative ions are repelled. The use of a conversion dynode is not necessary in a magnetic-sector mass spectrometer because the ions strike the detector with 3–10 kV of acceleration.

electron multiplier: (**continuous-** and **discrete-dynode EM**): An electron multiplier (EM) is a device used for current amplification through the secondary emission of electrons. Typically, the electron multiplier will produce 10^5 electrons ion^{-1} (a gain of 10^5). A **discrete-dynode EM** has a series of secondary emitters (dynodes). When an ion strikes the first dynode with sufficient velocity to produce the emission of a number of electrons (the emitting surfaces were constructed of a copper–beryllium material in earlier vintages), the electrons are attracted by the positively biased second dynode where even more electrons are produced. This process is repeated through several generations of secondary emissions. It is not uncommon for a discrete-dynode EM to have 10–20 stages. The typical operation voltages for this EM is 3–5 kV. The **continuous-dynode EM** uses a continuous surface of lead-doped glass that produces the same effect. The typical operating voltage of a continuous-dynode EM is <2 kV. Currently, there are also discrete-dynode electron multipliers that use the lead-doped glass and operate at a lower potential.

microchannel plate: (a.k.a. **channel electron multiplier**: **CEM**) The CEM is an array of 10^4 to 10^7 **continuous-dynode** electron multipliers (10–100 μm diam), side by side, each acting as an independent emitter to form a spatially resolved array detector. This type of detector is often used in conjunction with phosphor plates and a photodiode array to replace photographic plates in focal-plane mass spectrometers (mass spectrographs). They are also commonly found in TOF mass spectrometers.

Daly detector: In this detector, ions strike a dynode to produce secondary electrons that are directed toward a phosphor screen. When the electrons impact on the screen, photons are emitted and amplified with a photomultiplier tube. Unlike the EM or CEM, the amplification device is outside the vacuum system of the mass spectrometer. The phosphor screen experiences far less wear per ion impact than either the EM or CEM; therefore, this type of detector has a much longer life (several years as compared to 12–24 months or less for the EM and CEM).

Faraday cup: Positive ions arriving at a metal surface are neutralized by electrons that have passed through a high-ohm resistor. The ion current is calculated from the voltage drop across the resistor. A guard tip is required to prevent the loss of secondary electrons emitted by impacting ions. The measured and amplified ion current is directly proportional to the number of ions and to the number of charges per ion. The Faraday cup is an "absolute detector" and can be used to calibrate an electron multiplier. These detectors are free from mass discrimination because of a lack of dependence on ion velocity. This detector has a very long time delay that is inherent in its amplification system. It is used primarily for highly accurate measurements of slowly changing ion currents, such as the case with isotope ratio mass spectrometry. However, even with this response-time limitation, a detector has been proposed (based on the Faraday cup principle) that has an electronic signal amplification system fast enough to mass- and time-resolve MALDI TOFMS signals.[*]

---

[*] Bahr, U; Röhling, U; Lautz, C; Strupat, K; Schürenberg, M; Hillenkamp, F "A Charge Detector for TOF-MS of High-Mass Ions Produced by MALDI" *Int. J. Mass Spectrom. Ion Processes* **1996**, *153*, 9–22.

REFERENCED CITATIONS IN DOCUMENTS
(Reports, Journal Articles, Chapters, etc.)

Different journals and publishers have different styles for referenced articles and book chapters. They are usually contained in the Instructions to the Author section of journals. The names of the authors, journal name, year, volume number, and page number or page-number range are always required. The inclusion of the title of the referenced article or chapter is somewhat contentious. Although not the case with most modern biotechnology journals, many chemistry and mass spectrometry journals not only do not require titles in cited articles, but they also do not allow for them.

The overwhelming volume of information that is available in mass spectrometry (as well as all scientific areas) today requires a high degree of efficiency in researching a topic. Most users of mass spectrometry do not have sufficient time to review title listings or abstracts of articles in their field, let alone track every referenced citation in a research article. However, when a citation does not contain the title of an article, the determination of the value of one of these references often requires that the original article or chapter be consulted. The volume of information and the time demand on the researcher make burdensome the need to consult the original citation in the absence of its potential value. The lack of titles in the citations in review articles, such as those that appear from time to time in the Accounts and Perspective section of the *Journal of the American Society for Mass Spectrometry*, *Mass Spectrometry Reviews*, the Special Features section of the *Journal of Mass Spectrometry*, and the A-Pages of *Analytical Chemistry*, detracts from the utility of these articles.

These and other reasons have compelled many scientific research funding agencies such as the U.S. National Institutes of Health to require the inclusion of titles with all citations in grant proposals. *Mass Spectrometry Reviews* now requires the inclusion of titles; and the *Journal of the American Society for Mass Spectrometry* (Gross, Sparkman 1998), the *Journal of Mass Spectrometry*, and the *International Journal of Mass Spectrometry and Ion Processes* have just made editorial policy changes to allow (but not require [the policy is that if any cited reference includes a title, then all the citations within the article must include titles] consistency) for the inclusion of titles in cited references. The title of a scientific article is really a mini-abstract. When an article that contains titles in its references is reviewed, it is much easier to determine the potential value of retrieving one of these references.

The primary objection to the inclusion of titles in referenced citations is that many authors have extensive bibliographies without titles stored in their databases. The reason stated by many journals was the amount of space required; however, with the extensive amount of "white space" in most modern journals, this objection is without any reasonable merit.

It is hoped that future contributors to the scientific literature will see the value that titles add to references and will include them in citations, and that this practice becomes commonplace in submitted articles.

Note: The source for journal abbreviations is *The ACS Style Guide: A Manual for Authors and Editors*, 2nd ed.; pp 215–229 (Dodd 1997).

INTERNET CITATIONS IN PUBLISHED WORKS

As the Internet has become a more significant factor in the presentation of scientific information, some referenced information may only be available on the Internet. If information that is found not only on the Internet but also in a book or in a journal, then the book or journal should be cited. If the information is only available on the Internet, then the citation should have the appropriate and specific URL for the cited material in place of the journal name (or book title and publisher) and volume and page numbers. The date the material was posted should appear in the citation if possible. One of the problems with Web citations is the transient nature of the Web sites. There may come a time when material is considered permanent on the Web; but, for the present, Web citations should be considered much the same as private communications.

Example:

Sparkman, OD "Mass Spectrometry: A Primer" *spectroscopyNOW.com: BasePeak*; Wiley: Chichester, U.K.; http://www.spectroscopynow.com/Spy/basehtml/SpyH/1,1181, 4-14-9-0-0-education_dets-0-2552,00.html.

This HTML has a *Last Updated* statement of 06 October 2003 when this book was in preparation; however, the article first appeared on the site sometime in 2002. This is why referencing Web sources is considered to be incomplete. In this example, it would be possible to go to the top Web site of http://www.spectroscopynow.com and find the article; however, it is much better to include the specific URL. However, if your citation is not in an electronically accessible document, entering the URL can be somewhat cumbersome.

REQUIRED EXPERIMENTAL INFORMATION FOR PUBLICATIONS

The amount of information appearing in journal articles regarding the acquisition of data by GC/MS, LC/MS, and mass spectrometry is often insufficient for those wishing to reproduce the method to do so.

Mass spectrometry is increasingly being used by chromatographers who are used to treating whatever is on the exit end of the column as a detector. This means that a great deal of the mass spectrometry particulars necessary to reproduce an analysis are often omitted.

It is important to remember that whereas most GC/MS systems come with a specific gas chromatograph (GC) mated to a specific mass spectrometer, some manufacturers have different model GCs as options. Most LC/MS systems require a separate mass spectrometer (MS) and liquid chromatograph (LC) purchase; in some cases, the LC and MS are from two different manufacturers. In some cases, the LC/MS instrument is only a mass spectrometer with an "LC interface" and no liquid chromatograph. A good example of combinations of GCs and mass spectrometers that can be confusing is the case of the Agilent (then Hewlett-Packard) 5890 GC and 5972 MS followed by the 6890 GC and 5972 MS. There was also a 6820 GC with a 5973 MS and a 6890 GC with a 5973 MS.

As a general rule of thumb, any value that can be set in a method editor on a computerized data system or any value that can be read during acquisition should be reported.

The following is a list of required information:

Mass Spectrometer: manufacturer, **model**, and **type**; e.g., Varian, 4000, External Ionization Quadrupole Ion Trap, Walnut Creek, CA, USA.

> **Note:** Reporting the mass spectrometry instrument must include relevant details. An example of a potential lack of information in reporting just the manufacturer and model name (number) is seen for instruments that are available with various pumping system options. It is important to include the specifics of the pumping system with the instrument's description. The Agilent 5973 has been sold with an oil diffusion pump, a 70-L sec^{-1} turbomolecular pump, and a 250-L sec^{-1} turbomolecular pump. Both the 70-L sec^{-1} and the 250-L sec^{-1} turbomolecular pump instruments offer chemical ionization; however, CI performance on the 70-L sec^{-1} system is far inferior to the performance of the 250-L sec^{-1} system.
>
> Some manufacturers offer various types of electronic upgrades to a given model of an instrument. If such an upgrade has been installed on an instrument, the details should be included with the instrument's description.

> **type of sample introduction**; e.g., GC, LC, direct insertion probe, desorption CI or FD probe, flow injection, direct infusion, (etc.). Specifically, all makes and models of valves, syringe pumps, (etc.).

> **sign of ions/type of ionization**; e.g., positive EI, positive CI, FI, negative ECI, negative electrospray, positive liquid matrix secondary-ion mass spectrometry, (etc.).

> **ionization energy**; in the case of EI, CI, and ECI GC/MS (where a beam of electrons is directly or indirectly involved with the ionization), the electron energy should be reported; e.g., for an EI analysis: 70 eV. When ionization is a function of a laser, ion beam, beam of atoms, (etc.), the details of the source and level of energy and any matrix used in the

sample preparation should be reported; e.g., 10 keV Cs^+ as a primary ion-beam source and *m*-nitrobenzyl alcohol (*m*-NBA) as a matrix.

relevant voltages; the electrospray interface or needle voltage, the voltage that controls the velocity of ions between the atmospheric region and the *m/z* analyzer (cone voltage, fragmentation voltage, etc.), and so forth. Care must be taken to provide all the information required to reproduce the method on the specific make and model of the instrument used. The labels of these voltages vary from manufacturer to manufacturer. In the case of MALDI, the laser settings should be reported. In the case of an MS/MS analysis, the collision energy should be reported.

reagent gas and **pressure** (where appropriate); in the case of chemical ionization or electron capture/negative ionization where a reagent gas such as methane is used, the gas and its pressure in the ion source should be reported. If the pressure in the ion source cannot be determined, some determinable value should be reported such as the gauge pressure on the supply-size of the gas cylinder.

gases used; any gas such as a GC carrier gas, nebulizing gas, drying gas, curtain gas, or gas used with a particle beam interface should be reported as to type, source (gas generator, boil-off of a liquid, etc.), temperature, and flow rate and/or pressure. If an MS/MS experiment is being carried out, the collision gas and its pressure should be reported (i.e., Ar at a pressure of 3.0×10^{-1} Pa).

relevant pressures; e.g., pressure in the ion source, analyzer, interface in LC/MS systems, detector region of the analyzer, (etc.).

***m/z* scale calibration**; name of the manufacturer's tune file, the material used for calibration (e.g., PFTA, PFK, PPG, Agilent ESI tune-mix, etc.). In the case of an accurate mass measurement, the reference *m/z* value and compound should be reported. This can be very important for a MALDI analysis.

type of data acquisition; e.g., continuous acquisition of spectra or selected-ion monitoring (SIM), alternating continuous scan/SIM, MS/MS mode (fixed-precursor-ion analysis, fixed-product-ion analysis, or common-neutral-loss analysis), selected reaction monitoring (SRM), (etc.).

data acquisition range; e.g., *m/z* 20–350, SIM ions, SRM transitions, precursor ions, (etc.).

data acquisition rate; some instruments do not allow the direct setting of the acquisition rate (spectra per second or time for the acquisition of a single spectrum). These values are determined by setting the number of spectra to be averaged before saving a spectrum to a data file. For example, the Agilent MSD ChemStation for the 5973 allows the setting of the **start mass and end mass** on the **scanning mass range** tab of the **edit scan parameters** dialog box and the **sampling rate (2^n)** on the **threshold & sampling rates** tab. Changing the value in any of these three fields will change the value reported in the Scans/Sec field of the Summary Settings pane of the dialog box. The value entered in the **sampling rate (2^n)** is the power to which 2 is raised to determine the number of spectra that are averaged before a spectrum is stored. Both this number and the *data acquisition rate* (Scans/Sec) should be reported.

The Varian quadrupole ion trap mass spectrometer's data system has entry fields for the starting and ending *m/z* values for the data acquisition. It also has an entry field for the number of spectra to be acquired in one second. This causes the number of *microscans* (spectra averaged before a spectrum is stored) to change. Again, both the spectra per second and the number of microscans should be reported as the *data acquisition rate*.

In the case of an SIM or SRM analysis, the dwell time for each ion should be specified along with the time these ions are monitored during the analysis.

temperature of the ion source; e.g., 200 °C.

temperature of the analyzer (mass spectrometer); e.g., 150 °C.

other relevant temperatures; e.g., temperatures of the GC column transfer line, APCI interface (which is an ion-source temperature), ES interface, (etc.).

mass spectrometer's resolving power (or **resolution**); in the case of most GC/MS analyses using transmission quadrupole or QIT mass spectrometers, the resolution is unit (ions differing by 1 *m/z* unit can be separated) even though the data system can report to the nearest 0.05–0.01 *m/z* unit. This difference in resolution and reported value should be noted. In some cases, using a transmission quadrupole in the SIM mode, the instrument can be set to monitor the nearest 0.05 *m/z* unit. These used values should be reported in the acquisition range portion of the method.

In the case of data acquired under high-resolving-power conditions, all relevant information should be reported; e.g., reported resolving power and definition used (10,000 based on 10% valley definition or 20,000 based on FWHM definition), the reference compound (and *m/z* value used, if appropriate), the determined accuracy of mass measurements (usually in ppm).

type of ion detection device; e.g., electron multiplier, microchannel plate, photomultiplier, (etc.).

electron multiplier (or ion detection device) **voltage**; e.g., EM voltage = 1500. In the case of an electron multiplier, the gain should be reported even though, in most cases, it is 10^5.

solvent delay; the delay between the time the analysis was started (usually from the sample introduction time) and the time mass spectral acquisition begins. This is mainly used in GC/MS to make sure that the solvent elutes the column before the ionization filament and the EM are turned on.

data acquisition time; the amount of time that data are acquired.

Gas Chromatograph: manufacturer, model, and **type**; e.g., Agilent, 6890, 220 Volt, Little Falls, Delaware, USA.

injector type; e.g., split/splitless, cold-on-column, large volume, (etc.). All pressure used with the injector should be reported.

injection type (do not confuse with injector type); e.g., split, splitless, (etc.). If a split injection is used, the split ratio should be stated. If a splitless injection is being made, the split delay and the purge flow to split vent flow should be reported.

injector temperature; e.g., 250 °C.

GC column; e.g., column length (meters), column diameter (mm; i.e., 0.32 mm or µm; i.e., 320 µm, sometimes called 320 microns [sic]) station phase (DB™-5MS), thickness of stationary phase in open-tubular capillary columns (µm; 0.25 µm, sometimes called 0.25 microns [sic]), column's material of construction (fused silica).

carrier gas, **linear velocity**, **flow rate**; e.g., He, 32 cm sec^{-1}, 1 mL min^{-1}. There may be other parameters that are specific to an instrument make or an injector type.

column oven temperature details; initial temperature and hold time; rate of rise (°C min^{-1}) to next temperature; hold time for second plateau; rate of rise to next plateau, (etc.). If the initial temperature is a result of a cryogen, give particulars of the cryogen (CO_2 or LN_2) flow rates, (etc.).

other detectors; if other GC detectors such as the FID or ECD is used, report all relevant data such as temperature, gases, flow rate, (etc.).

Liquid Chromatograph: manufacturer, **model**, **and type**; e.g., Waters, ACQUITY, UPLC™ System, Milford, MA, USA.

LC column, **manufacturer**, **type**, **packing**, **i.d.**, **length**, and **size of particles**; e.g., Waters, SunFire C18 , 4.6mm × 100mm, 3.5 µm.

solvent A; e.g., composition 35% 20 mM K_2HPO_4 / 20 mM KH_2PO_4 (source of all materials).

solvent B; e.g., 65% methanol (source of all materials).

phase; e.g., normal or reversed phase.

separation conditions; e.g., isocratic or gradient elution. If gradient elution, report gradient.

mobile-phase flow rate; e.g., 1 mL min^{-1}.

analysis time; e.g., 100 min.

analysis temperature; e.g., ambient, 65 °C, (etc.).

other detectors; manufacturer, **make**, and **model** (**along with any conditions**); e.g., Waters, Variable Wavelength UV Detector, set at 214 nm.

Note: When a chromatographic device is used, the **sample size** and **method of introduction** should be specified. In the case of GC sample introduction, the use of slow or fast injections should be specified. If an **autosampler** is used, all settable parameters associated with the autosampler should be reported. The **sample solvent** should always be reported along with the exact or approximate **analyte concentration**.

Data Analysis: manufacturer, **software name**, **and version number**; e.g., Thermo Electron, Xcaliber™, Version 1.4, San Jose, CA, USA.

additional data analysis software including version number or publication date, **and where acquired**; e.g., MS Tools, v.5.14, purchased from ChemSW, Fairfield, CA, USA.

databases used in identifications including version number and where acquired; e.g., NIST Mass Spectral Database (NIST02), purchased from Scientific Instrument Services, Ringoes, NJ, USA.

REFERENCES

1. Abramson, FP "Chemical Reaction Interface Mass Spectrometry" *Mass Spectrom. Rev.* **1994**, *13*, 341–356.

2. Adams, J "Letter-To-Editor" *J. Am. Soc. Mass Spectrom.* **1992**, *3*, 473.

3. Aston, FW *Isotopes*; Edward Arnold: London, 1924.

4. Badman, ER; Wells, JM; Bui, HA; Cooks, RG "Fourier Transform Detection in a Cylindrical Quadrupole Ion Trap" *Anal. Chem.* **1998**, *70*, 3545–3547.

5. Barber, M; Bordoli, RS; Sedgwick, RD; Tyler, AN "Fast Atom Bombardment of Solids (F.A.B.): A New Ion Source for Mass Spectrometry" *J. Chem. Soc. Chem. Commun.* **1981**, 325–327.

6. Barnett, EF; Tandler, WSW; Turner, WR Quadrupole Mass Filter with Fringe Field Penetrating Structure. U.S. Patent 356,0734, 1971.

7. Barshick, CM "Glow Discharge Mass Spectrometry" in *Inorganic Mass Spectrometry: Fundamentals and Applications*; Barshick, CM; Duckworth, DC; Smith, DH, Eds.; Marcel Dekker: New York, 2000, ISBN:0824702433 (reviewed *JASMS* 11:822).

8. Bauer, S; Cooks, RG "MIMS for Trace Determination of Organic Analytes in On-line Process Monitoring and Environmental Analysis" *Am. Lab.* **1993**, 36–51.

9. Beavis, RC "Matrix-assisted Ultraviolet Laser Desorption: Evolution and Principles" *Org. Mass Spectrom.* **1992**, *27*, 653–659.

10. Becker, EW; Bier, K; Burghoff, H; Zigan, F "The Separation Jet" *Z. Naturfosch. Tell A* **1957**, *12*, 609.

11. Beckey, HD *Principles of Field Ionization and Field Desorption Mass Spectrometry*; Pergamon: New York, 1977.

12. Beckey, HD Field Ionization Mass Spectrometry. In *Advances in Mass Spectrometry*, Vol. 2; Elliott, RM, Ed.; Pergamon: Oxford, U.K., 1963; p 1.

13. Beynon, JH *Mass Spectrometry and Its Applications to Organic Chemistry*; Elsevier: Amsterdam, 1960 (reprinted by ASMS, 1999).

14. Beynon, JH; Zerbi, G "IUPAC Recommendations on Symbolism and Nomenclature for Mass Spectrometry" *Org. Mass Spectrom.* **1977**, *12*(3), 115–118.

15. Beynon, JH "Report of Nomenclature Committee Workshop" *Proceedings of the 29th ASMS Conference on Mass Spectrometry and Allied Topics*, Minneapolis, Minnesota, May 24–29, 1981; pp 797–815.

16. Biemann, K *Mass Spectrometry: Organic Chemical Applications*, Vol. 1; McGraw-Hill: New York, 1962 (reprinted as *ASMS Publications Classic Works in Mass Spectrometry*; American Society for Mass Spectrometry: Santa Fe, NM, 1998).

17. Blakely, CR; Vestal, ML "Thermospray Interface for Liquid Chromatography/Mass Spectrometry" *Anal. Chem.* **1983**, *55*, 750–754.

18. Brubaker, WM Quadrupole MS. In *Advances in Mass Spectrometry,* Vol. 4; Kendrick, E, Ed.; Institute of Petroleum: London, 1968; p 293.

19. Bruins, AP; Covey, TR; Henion, JD "Ionspray Interface for Combined Liquid Chromatography/Atmospheric Pressure Ionization Mass Spectrometry" *Anal. Chem.* **1987**, *59*, 2642–2646.

20. Budde, WL *Analytical Mass Spectrometry: Strategies for Environmental and Related Applications*; American Chemical Society: Washington, DC; Oxford: New York, 2001.

21. Budzikiewicz, H; Djerassi, C; Williams, DH *Interpretation of Mass Spectra of Organic Compounds*; Holden-Day: San Francisco, CA, 1964.

22. Bursey, MM; Hoffman, MK Mechanisms Studies of Fragmentation Pathways. In *Mass Spectrometry: Techniques and Applications*; Milne, GWA, Ed.; Wiley–Interscience: New York, 1971.

23. Bursey, MM "Editorial" *Mass Spectrom. Rev.* **1992**, *11*, 1, 2.

24. Bursey, MM "Editorial" *Mass Spectrom. Rev.* **1991**, *10*, 1, 2.

25. Buryakov, IA; Krylov, EV; Nazarov, EG; Rasulev, UK "A New Method of Separation of Multi-Atomic Ions by Mobility at Atmospheric Pressure Using A High-Frequency Amplitude-Asymmetric Strong Electric Field" *Int. J. Mass Spectrom. Ion Processes* **1993**, *128*, 143–148.

26. Buryakov, IA; Krylov, EV; Makas, AL; Nazarov, EG; Pervukhin, VV; Rasulev, UK "Separation of Ions According to Mobility in Strong AC Fields" *Sov. Tech. Phys. Lett.* **1991**, *17*(6), 446, 447.

27. Busch, KL "Electron Ionization, Up Close and Personal" *Spectroscopy* **1995**, *10*(8), 39–42.

28. Caprioli, RM; Fan, T; Cottrell, JS "Continuous-flow Sample Probe for Fast Atom Bombardment Mass Spectrometry" *Anal. Chem.* **1986**, *58*(14), 2949–2953.

29. Chadwick, J "Possible Existence of the Neutron" *Nature* **1932**, *129*, 312.

30. Chargaff, E "Chemical Specificity of Nucleic Acids and Mechanism of Their Enzymatic Degradation" *Experentia* **1950**, *6*, 201–209.

31. Cody, RB; Freiser, BS "Electron Impact Excitation of Ions from Organics: An Alternative to Collision Induced Dissociation" *Anal. Chem.* **1979**, *51*, 547–554.

32. Cody, RB; Laramée, JA; Durst, HD "A Versatile New Ion Source for the Analysis of Materials in Open Air Under Ambient Conditions" *Anal. Chem.* **2005**, *77*, 2297–2302.

33. Cooks, RG; Beynon, JH; Caprioli, RM; Lester, GR *Metastable Ions*; Elsevier: New York, 1973.

34. Cooks, RG; Rockwood, AL "Letter-To-Editor" *Rapid Commun. Mass Spectrom.* **1991**, *5*(2), 93.

35. Cooks, RG "Frontiers in Quadrupole Ion Traps" *Proceedings of the 51st ASMS Conference on Mass Spectrometry and Allied Topics*, Montreal, Quebec, Canada, June 8–12, 2003.

36. Cooper, H; Hakånsson, K; Marshall, AG "The Role of Electron Capture Dissociation in Biomolecular Analysis" **2005**, *24*, 201–222.

37. de Hoffmann, E; Charette, J; Stroobant, V Appendix 1: Nomenclature; Appendix 2: Abbreviations. *Mass Spectrometry: Principles and Applications*; Wiley: Chichester, U.K., 1996.

38. de Hoffmann, E "Tandem Mass Spectrometry: A Primer; Special Feature: Tutorial" *J. Mass Spectrom.* **1996**, *31*(2), 129–137.

39. Dempster, AJ "Positive Ray Analysis of Potassium, Calcium and Zinc" *Phys. Rev.* **1922**, *20*, 631–638.

40. Denoyer, E; Van Grieken, R; Adams, F; Natusch, DFS "Laser Microprobe Mass Spectrometry 1: Basic Principles and Performance Characteristics" *Anal. Chem.* **1982**, *54*(1), 26A–41A.

41. Dodd, JS, Ed. *The ACS Style Guide*; American Chemical Society: Washington, DC, 1986.

42. Dodd, JS, Ed. *The ACS Style Guide: A Manual for Authors and Editors*, 2nd ed.; American Chemical Society: Washington, DC, 1997.

43. Dougherty, RC; Dalton, J; Biros, FJ "Negative Ionization of Chlorinated Insecticides" *Org. Mass Spectrom.* **1972**, *6*, 1171–1181.

44. Dunbar, RC "BIRD (Blackbody Infrared Radiative Dissociation): Evolution, Principles, and Applications" *Mass Spectrom. Reviews* **2004**, *23*, 127–158.

45. Emmett, MR; White, FM; Hendrickson, CL; Shi, SD-H; Marshall, AG "Application of Micro-Electrospray Liquid Chromatography Techniques to FT-ICR MS to Enable High-Sensitivity Biological Analysis" *J. Am. Soc. Mass Spectrom.* **1997**, *9*, 333.

46. Emmett, MR; Caprioli, RM "Release of Endogenous Methionine Enkephalin in Brain Using Microdialysis and Capillary LC, Micro-ES/MS/MS" *Proceedings of the 42nd ASMS Conference on Mass Spectrometry and Allied Topics*, Chicago, Illinois, May 29–June 3, 1994; p 420.

47. (2)Emmett, MR; Caprioli, RM "Micro-Electrospray Mass Spectrometry: Ultra-High-Sensitivity Analysis of Peptides and Proteins" *J. Am. Soc. Mass Spectrom.* **1994**, *5*, 605.

48. Emmett, MR; Caprioli, RM "Nanoliter Flow-LC/ES/MS for the Analysis of Endogenous Neuropeptides" *Proceedings of the 41st ASMS Conference on Mass Spectrometry and Allied Topics*, San Francisco, California, May 31–June 4, 1993; pp 47a, 47b.

49. Falkner, FC "Letter-To-Editor" *Biomed. Environ. Mass Spectrom.* **1977**, *4*(1), 66, 67.

50. Field, FH Chemical Ionization Mass Spectrometry. In *Mass Spectrometry*, Vol. 5; Maccoll, A, Ed.; MTP Int. Rev. Sci. Butterworths: London, 1972; pp 133–138.

51. FitzGerald, GF "Dissociation of Atoms" *The Electrician* **1897**, *39*, 103, 104.

52. Freiser, B "Gas-phase Metal Ion Chemistry" *J. Mass Spectrom.* **1996**, *31*, 703–715.

53. Gamow, G *Thirty Years that Shook Physics: the Story of Quantum Theory*; (translated from German by Barbara Gamow); Doubleday: New York, 1966.

54. Ghaderi, S; Kulkarni, PS; Ledford, EB, Jr.; Wilkins, CL; Gross, ML "Chemical Ionization in Fourier Transform Mass Spectrometry" *Anal. Chem.* **1981**, *53*, 428–437.

55. Glish, G "Letter-To-Editor" *J. Am. Soc. Mass Spectrom.* **1992**, *2*, 349.

56. Gold, V "IUPAC Glossary of Terms Used in Physical Organic Chemistry" *Pure Appl. Chem.* **1983**, *55*, 1281; specific definitions on pp 1296, 1297.

57. Gomer, R; Inghram, MG "Letters to the Editor: Mass Spectrometric Analysis of Ions from the Field Microscope" *J. Chem. Phys.* **1954**, *22*, 1279; "Communication to the Editor: Applications of Field Ionization to Mass Spectrometry" *J. Am. Chem. Soc.* **1955**, *77*, 500.

58. Gross, ML "Editorial" *J. Am. Soc. Mass Spectrom.* **1994**, *5*, 57.

59. Gross, ML; Sparkman, OD "Editorial" *J. Am. Soc. Mass Spectrom.* **1998**, *9*, 451.

60. Gygi, SP; Rist, B; Gerber, SA; Tureček, F; Gelb, MH; Aebersold, R "Quantitative Analysis of Complex Protein Mixtures Using Isotope-Coded Affinity Tags" *Nature Biotechnology* **1999**, *17*, 994–999.

61. Haddon, WF Organic Trace Analysis Using Direct Probe Sample Introduction and High Resolution Mass Spectrometry. In *High Performance Mass Spectrometry: Chemical Applications*; Gross, ML, Ed.; ACS Symposium Series 70; Gould, RF, Ed.; American Chemical Society: Washington, DC, 1978; pp 97–119.

62. Hager, JW "A New Linear Ion Trap Mass Spectrometer" *Rapid Commun. Mass Spectrom.* **2002**, *16*, 512–526.

63. Hammer, C-G; Holmstedt, B; Ryhage, R "Mass Fragmentography of Chlorpromazine" *Anal. Biochem.* **1968**, *25*, 532–548.

64. Harrison, WW; Hess, KR; Marcus, RK; King, FL "Glow Discharge Mass Spectrometry" *Anal. Chem.* **1986**, *58*, 341A–356A.

65. Hertz, HS; Hites, RA; Biemann, K "Identification of Mass Spectra by Computer Searching a File of Known Spectra" *Anal. Chem.* **1971**, *43*, 681–691.

66. Hill, EA "On a System of Indexing Chemical Literature; Adopted by the Classification Division of the U.S. Patent Office" (a read before the Washington Section of the American Chemical Society, May 10, 1900) *J. Am. Chem. Soc.* **1900**, *22*(8), 478–494.

67. Hill, HH, Jr.; Siems, WF; St. Louis, RH; McMinn, DG "Ion Mobility Spectrometry" *Anal. Chem.* **1990**, *62*(23), 1201A–1209A.

68. Hillenkamp, F; Karas, M; Beavis, RC; Chait, BT "Matrix-Assisted Laser Desorption/Ionization Mass Spectrometry of Biopolymers" *Anal. Chem.* **1991**, *63*, 1193A–2003A.

69. Hillenkamp, F; Unsöld, E; Kaufmann, R; Nitsche, R "A High-Sensitivity Laser Microprobe Mass Analyzer" *Appl. Phys.* **1975**, *8*, 341–348.

70. Hirschfeld, T "The Hy-phen-ated Methods" *Anal. Chem.* **1980**, *52*, 297A–312A.

71. Hites, RA, Biemann, K "Computer Evaluation of Continuously Scanned Mass Spectra of Gas Chromatographic Effluents" *Anal. Chem.* **1970**, *42*, 855–860.

72. Horning, EC; Horning, MG; Carroll, DI; Dzidic, I; Stillwell, RN "New Picogram Detection System Based on Mass Spectrometry with an External Ionization Source at Atmospheric Pressure" *Anal. Chem.* **1973**, *45*, 936–943.

73. Hu, Q; Noll, RJ; Li, H; Makarov, A; Hardmanc, M; Cooks, RG "The Orbitrap: A New Mass Spectrometer" *J. Mass Spectrom.* **2005**, *40*, 430–433.

74. Hunt, DF; Stafford, GC, Jr.; Crow, FW; Russell, JW "Pulsed Positive Negative Chemical Ionization Mass Spectrometry" *Anal. Chem.* **1976**, *48*, 2098–2105.

75. Hutchens, TW; Yip, T-T "New Desorption Strategies for the Mass Spectrometric Analysis of Macromolecules" *Rapid Comm. Mass Spectrom.* **1993**, *7*, 576–580.

76. Hutchens, TW; Yip, T-T "Differences in Conformational States of a Zinc-finger DNA-binding Protein Domain Occupied by Zinc and Copper Revealed by Electrospray Ionization Mass Spectrometry" *Rapid Comm. Mass Spectrom.* **1992**, *6*, 469–473.

77. Ito, Y; Takeuchi, T; Ishii, D; Goto, M "Direct Coupling of Micro High-performance Liquid Chromatography with Fast Atom Bombardment Mass Spectrometry" *J. Chromatogr.* **1985**, *346*, 161–166.

78. Karas, M; Bachmann, D; Bahr, U; Hillenkamp, F "Matrix-assisted Ultraviolet Laser Desorption of Non-volatile Compounds" *Int. J. Mass Spectrom. Ion Processes* **1987**, *78*, 53–68.

79. Karas, M; Bachmann, D; Hillenkamp, F "Matrix Assisted UV Laser Desorption of Non-volatile Organic Compounds" in *Advances in Mass Spectrometry*, Vol. 10B; Todd, JFJ, Ed.; Wiley: Chichester, 1986; p 969.

80. Karasek, FW; Clement, RE Gas Chromatography-Mass Spectrometry Glossary. In *Basic Gas Chromatography-Mass Spectrometry Principles And Techniques*; Elsevier Scientific: Amsterdam, 1981.

81. Karni, M; Mandelbaum, A "The Even-Electron Rule" *Org. Mass Spectrom.* **1980**, *15*(2), 53–64.

82. Keenan, CW; Wood, JH *General College Chemistry*, 2nd ed.; Harper & Brothers: New York, 1961; 1st ed. 1957.

83. Kelly, PE; Stafford, GC "Performance Characteristics and Analytical Applications of Advanced Ion Trap Mass Spectrometers" *Proceedings of the 31st ASMS Conference on Mass Spectrometry and Allied Topics*, Boston, MA, May 8–13, 1983; p 530.

84. Kingdon, KH "A Method for Neutralization of Electron Space Charge by Positive Ionization at Very Low Gas Pressures" *Phys. Rev.* **1923**, *21*, 48.

85. Kiser, RW *Introduction to Mass Spectrometry and Its Applications*; Prentice-Hall: Englewood Cliffs, NJ, 1965; Introduction, p 1.

86. Knight, RD "Storage of Ions from Laser-Produced Plasma" *Appl. Phys. Lett.* **1981** *38*, 221.

87. Kondrat, RW; Cooks, RG "Direct Analysis of Mixtures by Mass Spectrometry" *Anal. Chem.* **1978**, *50*, 81A–92A.

88. Kuehl, DW "Identification of Trace Contaminants in Environmental Samples by Selected Ion Summation Analysis of Gas Chromatographic-Mass Spectral Data" *Anal. Chem.* **1977**, *49*, 521, 522.   Budzikiewicz, H; Djerassi, C; Williams, DH *Interpretation of Mass Spectra of Organic Compounds*; Holden-Day: San Francisco, CA, 1964.

89. Larameé, JA; Cody, RB; Deinzer, ML "Discrete Energy Electron Capture Negative Ion Mass Spectrometry" in *Encyclopedia of Analytical Chemistry*, Vol. 13; Sparkman, OD, Section Ed.; Meyers, RA, Ed.; Wiley: Chichester, U.K., 2000, pp 11651–11679.

90. Laskin, J; Futrell, JH "Activation of Large Ions in FT-ICR Mass Spectrometry" *Mass Spectrom. Reviews* **2005**, *24*, 135–167.

91. Lehmann, WD "Pictograms for Experimental Parameters in Mass Spectrometry" *J. Am. Soc. Mass Spectrom.* **1997**, *8*, 756–759.

92. Llewellyn, PM; Littlejohn, DP in *Abstracts of the 11th Annual Conference on Analytical Chemistry and Applied Spectroscopy*, Pittsburgh, Pennsylvania, February 1966.

93. Makarov, A Mass Spectrometer. U.S. Patent 5,886,346, 1999.

94. Makarov, A "Electrostatic Axially Harmonic Orbital Trapping: A High-Performance Technique of Mass Analysis" *Anal. Chem.* **2000**, *72*, 1156.

95. March, RE; Todd, JFJ, Eds. *Practical Aspects of Ion Trap Mass Spectrometry*, Volume II *Ion Trap Instrumentation*; CRC: Boca Raton, FL, 1995.

96. Markey, SP *July 2000 Private Communication* based on: Randall, EW; Yoder, CH; Zuckerman, JJ. "Nuclear Magnetic Resonance Studies of Exchange in Stannylamines" *J. Am. Chem. Soc.* **1967**, *89*(14), 3438–3441.

97. Markides, K; Gräslund, A "Advanced Information on the Nobel Prize in Chemistry 2002, 9 October 2002" *Kunge Vetenskapsakademien*; The Royal Swedish Academy of Sciences: Stockholm, Sweden.
(http://www.nobel.se/chemistry/laureates/2002/chemadv02.pdf)

98. McLafferty, FW *Interpretation of Mass Spectra*, 2nd ed.; Benjamin/Cummings: Redding, MA, 1973.

99. McLafferty, FW; Hertel, RH; Villwock, RD "Probability Based Matching of Mass Spectra" *Org. Mass Spectrom.* **1974**, *9*, 690–702.

100. McLafferty, FW; Bockhoff, FM "Separation/Identification Systems for Complex Mixtures Using Mass Separation and Mass Spectral Characterization" *Anal. Chem.* **1978**, *50*, 69–78.

101. McLafferty, FW, Ed. *Tandem Mass Spectrometry*; Wiley–Interscience: New York, 1983.

102. McLafferty, FW; Tureček, F Glossary and Abbreviations. *Interpretation of Mass Spectra*, 4th ed.; University Science: Mill Valley, CA, 1993.

103. McLafferty, FW; Tureček, F *Interpretation of Mass Spectra*, 4th ed.; University Science: Mill Valley, CA, 1993; p 192.

104. Meng, CK; Mann, M; Fenn, JB "Of Protons or Proteins: A Beam's Beam" *Z. Phys.* **1988**, *D10,* 361–368.

105. Micromass *Back to Basics*, CD01, Version 2, Manchester, U.K., Dec. 2000.

106. Millikan, RA "A New Modification of the Cloud Method of Measuring the Elementary Electrical Charge and the Most Probable Value of that Charge" *Phys. Rev.* **1909**, *29*, 560, 561.

107. Mills, I; Cvitaš, T; Homann, K; Kallay, N; Kuchitsu, K *Quantities, Units and Symbols in Physical Chemistry*, 2nd ed.; International Union of Pure and Applied Physical Chemistry Division; Blackwell Science: London, 1993.

108. Mislow, K; Siegel, J "Stereoisomerism and Local Chirality" *J. Am. Chem. Soc.* **1984**, 106, 3319–3328.

109. Mulder, GJ "On the Composition of some Animal Substances*" J. für Parktische Chemie* **1839**, *16*, 129 (a partial translation and excerpted version can be found in *A Documentary History of Biochemistry 1770–1940* by Mikulás Teich, Rutherford, NJ; Fairleigh Dickinson University Press: 1992; and on the Web at http://webserver.lemoyne.edu/faculty/giunta/mulder.html).

110. Munson, B; Field, FH "Chemical Ionization Mass Spectrometry" *J. Am. Chem. Soc.* **1966**, *88*, 2621–2630; Munson, B "CI-MS: 10 Years Later" *Anal. Chem.* **1977**, *49*, 772A–778A.

111. Nelson, RW; Krone, JR; Bieber, AL; Williams, P "Mass Spectrometric Immunoassay" *Anal. Chem.* **1995**, *67*, 1153–1158.

112. Nicholson, AJC "Photochemical Decomposition of Aliphatic Methyl Ketones" *Trans. Faraday Soc.* **1954**, *50*, 1067–1073.

113. O'Farrell, PH "High Resolution Two-Dimensional Electrophoresis of Proteins" *J. Biol. Chem.* **1975**, 250, 4007–4021.

114. Olah, GA "Five-coordinated Carbon: Key to Electrophilic Reactions" *CHEMTECH* **1971**, 1, 566.

115. Olah, GA "The General Concept of Carbocations Based on Differentiation of the Trivalent ('Classical') Carbenium Ions from Three-center Bound Penta- or Tetracoordinated ('Nonclassical') Carbonium Ions: The Role of Carbocations in Electrophilic Reactions" *J. Am. Chem. Soc.* **1972**, *94*, 808.

116. Price, P "Standard Definition of Terms Relating to Mass Spectrometry" *J. Am. Soc. Mass Spectrom.* **1991**, *2*, 336–348.

117. Purves, RW; Guevremont, R; Day, S; Pipich, CW; Matyjaszczyk, MS "Mass Spectrometric Characterization of a High-Field Asymmetric Waveform Ion Mobility Spectrometer" *Review of Scientific Instruments* **1998**, *69*(12), 4094–4106.

118. Radom, L; Bouma, WJ; Nobes, RH; Yates, BF "A Theoretical Approach to Gas-Phase Ion Chemistry" *Pure Appl. Chem.* **1984**, *56*, 1831.

119. Rosenstock, HR; Melton, CE "Metastable Transionic and Collision-induced Dissociations in Mass Spectra" *J. Chem. Phys.* **1957**, *26*, 1303–1309.

120. Rutherford, E "Nuclear Constitution of Atoms" *Proc. R. Soc. London* **1920**, *97*, 374–400.

121. Ryhage, R "Mass Spectrometry as a Detector for GC" *Anal. Chem.* **1964**, *36*, 759–764.

122. Sakamoto, S; Fujita, M; Kim, K; Yamaguchi, K "Characterization of Self-Assembling Nano-Sized Structures by Means of Coldspray Ionization Mass Spectrometry" *Tetrahedron* **2000**, *56*, 955–964.

123. Scalf, M; Westphall, MS; Smith, LM "Charge Reduction Electrospray Mass Spectrometry" *ASMS 11th Sanibel Conference on Mass Spectrometry/Mass Spectrometry in Clinical Diagnosis of Disease*, Sanibel Island, Florida, January 23–26, 1999.

124. Schwartz, JC; Senko, MW; Syka, JEP "A Two-dimensional Quadrupole Ion Trap Mass Spectrometer" *J. Am. Soc. Mass Spectrom.* **2002**, *13*, 659–669.

125. Senko, MW; Canterbury, JD; Guan, S; Marshall, AG "A High-Performance Modular Data System for Fourier Transform Ion Cyclotron Resonance Mass Spectrometry" *Rapid Comm. Mass Spectrom.* **1996**, *10*(14), 1839–1844.

126. Smith, D; Španěl, P "Selected Ion Flow Tube Mass Spectrometry (SIFTMS) for On-Line Trace Gas Analysis" *Mass Spectrom. Rev.* **2005**, *24*, 661–700.

127. Sokolow, S; Karnofsky, J; Gustafson, P *Finnigan Application Report No. 2*; Finnigan: San Jose, CA, March 1978.

128. Stein, S, et al. "NIST/EPA/NIH Mass Spectral Database" and "NIST98 Mass Spectral Search Program, v 1.6" Mass Spectrometry Data Center, National Institute of Standards and Technology, U.S. Department of Commerce, Gaithersburg, MD; source of mass spectra.

129. Stephenson, JL, Jr.; McLuckey, SA "Charge Reduction of Oligonucleotide Anions Via Gas-phase Electron Transfer to Xenon Cations" *Rapid Commun. Mass Spectrom.* **1997**, *11*(8), 875–880.

130. Stevens, ED "Letter to the Editor" *Am. Lab.* **2001**(May), *33*(10), 8.

131. Stevenson, DP "Ionization and Dissociation by Electronic Impact, The Ionization Potentials and Energies of Formation of *sec*-Propyl and *tert*-Butyl Radicals, Some Limitations on the Method" *Discuss. Faraday Soc.* **1951**, *10*, 35.

132. Stoney, GJ "On Physical Units of Nature" *Philos. Mag.* **1881**, *11*, 81–90.

133. Stoney, GJ "On the Cause of Double Lines and of Equidistant Satellites of the Spectra of Gases" *J. Sci. Trans. R. Dublin Soc.* **1891**, [2]*4*, 563–608.

134. Stoney, GJ "On the 'Electron' or Atom of Electricity" *Philos. Mag.* **1894**, *38*, 418.

135. Strong, FC, III; Howard, B "Comments on Spectroscopic Nomenclature" *Appl. Spectrosc.* **1975**, *29*(3), 265, 266.

136. Strong, FC, III; Howard, B "Analysis versus Determination" *Am. Lab.* **2001-1** (January), *33*(1), 10.

137. Strong, FC, III; Howard, B "Letter to the Editor" *Am. Lab.* **2001-2**(August), *33*(16), 8.

138. Sundqvist, B; Macfarlane, RD "252-Cf Plasma Desorption Mass Spectrometry" *Mass Spectrom. Rev.* **1985**, *4*, 421–460.

139. Sweeley, CC; Elliott, WH; Fries, I; Ryhage, R "Mass Spectrometry Determination of Unresolved Components in Gas Chromatography" *Anal. Chem.* **1966**, *38*, 1549–1553.

140. Syka, JEP; Coon JJ; Schroeder, MJ; Shabanowitz, J; Hunt, DF "Peptide and protein sequence analysis by electron transfer dissociation mass spectrometry" *Proc. Natl. Acad. Sci. U.S.A.* **2004**, *101*(26) 9528–9533.

141. Takáts, Z; Wiseman, JM; Gologan, B; Cooks, RG "Mass Spectrometry Sampling Under Ambient Conditions with Desorption Electrospray Ionization" *Science* **2004**, *306*, 471–473.

142. Tal'roze, VL; Lyubimova, AK "Secondary Processes in a Mass Spectrometer Ion Source" *Doki Akad. Nauk SSSR* **1952**, *86*, 909.

143. Tanaka, K; Ido, Y; Akita, S; Yoshida, Y; Yoshida, T "Detection of High Mass Molecules by Laser Desorption Time-of-flight Mass Spectrometry" *Proceedings of the Second Japan–China Joint Symposium on Mass Spectrometry,* September 15–18, 1987, Osaka, Japan; Matsuda, H; Xiao-tian, L, Eds.; Bando: Osaka, Japan, 1987.

144. Tanaka, K; Waki, H; Ido, Y; Akita, S; Yoshida, Y "Protein and Polymer Analysis up to *m/z* 100,000 by Laser Ionization Time-of-flight Mass Spectrometry" *Rapid Commun. Mass Spectrom.* **1988**, *2*(8),151–153.

145. Thomson, JJ "Cathode Rays" *The Electrician* **1897**, *39*, 104–122.

146. Thomson, JJ "Cathode Rays" *Philos. Mag.* **1897**, *44*, 293–316.

147. Todd, JFJ International Union of Pure and Applied Chemistry, Physical Chemistry Division, Commission on Molecular Structure and Spectroscopy [sic], Subcommittee on Mass Spectroscopy [sic], "Recommendations For Nomenclature And Symbolism For Mass Spectroscopy" [sic] (including an appendix of terms used in vacuum technology) (*Current IUPAC Recommendations*, 1991); Todd, JFJ *Int. J. of Mass Spectrom. Ion Processes* **1995**, *142*(3), 211–240.

148. U.S. Government *United States Government Printing Office Manual-Of-Style*; United States Government Printing Office: Washington, DC, 1986; p 134.

149. Wasinger, V; Cordwell, SJ; Cerpa-Poljak, A; Gooley, AA; Wilkins, MR; Duncan, M; Williams, KL; Humphery-Smith, I "Progress With Gene-Product Mapping of the Mollicutes: Mycoplasma Genitalium" *Electrophoresis* **1995**, *16*, 1090–1094.

150. Watson, JT *Introduction to Mass Spectrometry*, 3rd ed.; Lippincott-Raven: Philadelphia-New York, 1997; p 118.

151. Watson, JT; Falkner, FC; Sweetman, BJ "Letter-To-Editor" *Biomed. Mass Spectrom.* **1974**, *1*, 156, 157.

152. Watson, JT; Biemann, K "High-resolution Mass Spectrometry with Gas Chromatography" *Anal. Chem.* **1964**, *36*, 1135–1137.

153. Wechsung, R; Hillenkamp, F; Kaufmann, R; Nitsche, R, Unsöld, E; Vogt, H LAMMA: A New Laser Microprobe Mass Analyzer. In *Microscopica Acta, Supplement 2: Microprobe Analysis in Biology and Medicine*; Springer-Verlag: Stuttgart, Germany, 1978; pp 281–296.

154. Wei, J; Buriak, JM; Siuzdak, G "Desorption-ionization Mass Spectrometry on Porous Silicon" *Nature* **1999**, *399*, 241–244.

155. Whitehouse, CM; Fenn, JB; Yamashita, M "An Electrospray Ion Source for Mass Spectrometry of Fragile Organic Species" *Proceedings of the 32nd ASMS Conference on Mass Spectrometry and Allied Topics*, San Antonio, Texas, May 27–June 1, 1984; pp 188, 189.

156. Wilkins, MR; Sanchez, J-C; Gooley, AA; Appel, RD; Humphery-Smith, I; Hochstrasser, DF; Williams, KL "Progress with Proteome Projects: Why All Proteins Expressed By a Genome Should be Identified and How To Do It" *Biotechnology and Genetic Engineering Reviews* **1995**, *13*, 19–50.

157. Willoughby, RC; Browner, RF "Monodisperse Aerosol Generation Interface for LC/MS" *Anal. Chem.* **1984**, *56*, 2626.

158. Wilm, M; Mann, M "Analytical Properties of the Nanoelectrospray Ion Source" *Anal. Chem.* **1996**, *68*, 1.

159. Wilm, M; Mann, M "Micro Electrospray Source for Generating Highly Resolved MS/MS Spectra on 1 μL Sample Volume" *Proceedings of the 42nd ASMS Conference on Mass Spectrometry and Allied Topics*, Chicago, Illinois, May 29–June 3, 1994; p 420.

160. Winkler, H *Verbreitung und Ursache der Parthenogenesis im Pflanzenund Tierreiche*; Verlag Fischer: Jena, Germany, 1920; p 165.

161. Wise, BM "What is Chemometrics" *ASMS 13th Sanibel Conference on Mass Spectrometry in Informatics and Mass Spectrometry*, Sanibel Island, Florida, January 20–22, 2001.

162. Yates, BF; Bouma, WJ; Radom, L "Detection of the Prototype Phosphonium (CH_2PH_3), Sulfonium (CH_2SH_2), and Chloronium (CH_2C1H) Ylides by Neutralization–Reionization Mass Spectrometry: A Theoretical Prediction" *J. Am. Chem. Soc.* **1984**, *106*, 5805–5808.

163. Yergey, J; Heller, D; Hansen, G; Cotter, RJ; Fenselau, C "Isotope Dilution in MS of Large Molecules" *Anal. Chem.* **1983**, *55*, 353–356.

164. Yost, RA; Enke, CG; McGilvery, DC; Smith, D; Morrison, JD "High Efficiency Collision-Induced Dissociation in an RF-Only Quadrupole" *Int. J. Mass Spectrom. Ion Phys.* **1978**, *30*, 127–136.

165. Zubarev, RA; Kelleher, NL; McLafferty, FW "Electron Capture Dissociation of Multiply Charged Protein Cations. A Nonergodic Process" *J. Am. Chem. Soc.* **1998**, *120*, 3265–3266.

BIBLIOGRAPHY

Books are important sources of information for the use of mass spectrometry. Books give an overview of a topic and often contain many seminal references to specific techniques. Unfortunately, like many of the chemistry-oriented journal articles (this is not true for the journals and books in the biological sciences), citations in mass spectrometry books often only contain bibliographical information—not the titles of journal articles. When a book such as Watson's *Introduction to Mass Spectrometry* (Introductory 11) that is filled with relevant references to journal articles comes along, it becomes a very coveted addition to a personal library because of the increased usefulness of these titles.

According to an *Analytical Chemistry* A-Pages article (Braun, T; Schubert, A; Schubert, G "The Most Cited Books in Analytical Chemistry" *Anal. Chem.* **2001**, *73*(23), A-667, 668), 10% of references in the hard sciences (chemistry and physics) were to books during the 1950s and 1960s; the remaining were to journal articles. This compared to 50% for the social sciences. In the 20-year period from 1980 to 1999, this citation of books compared with journal articles had dropped to ~3%. Of the 25 most often cited books over this period, 3 were related in some way to mass spectrometry. The two most often cited mass spectrometry titles were those by McLafferty (Interpretation 7) and Harrison (Technique 66) with about an equal number of citations (58 and 57, respectively). The first five books in this article were the only ones with triple-digit citations; and, of those, only *Electrochemical Methods: Fundamentals and Applications* (Brad, AJ; Faulkner, LR; Wiley: New York, 1980; 2nd ed., 2000) had more than 200 citations (374). However, the second book in this list, with 167 citations, has relevance to the mass spectrometry of today. This book is Snyder and Kirkland's classic on liquid chromatography (Introductory 26), first published in 1974, with the second edition appearing in 1979. This book is still in print. The article also stated that as of February 16, 2001, *A Global View of LC/MS: How to Solve Your Most Challenging Analytical Problems* (Technique 15) was the fourth "bestseller" according to www.amazon.com. In their survey of the bestsellers from a New Zealand Web site, the first was McLafferty and Tureček's *Interpretation of Mass Spectra* (Interpretation 7); followed by Snyder, Kirkland, and Glajch's *Practical HPLC Method Development* (Reference 23), second; and Kitson's *Gas Chromatography and Mass Spectrometry: A Practical Guide* (Technique 52), third. The authors pointed out that the relative low number of citations for books painted a picture far from the true importance of books. Quoting from "The Dynamic Scientific-information User" in *Communication: The Essence of Science* (Garvey, WD; Tomita, K; Woolf, P; Oxford: Pergamon, 1979; Appendix H), the importance of the book to science was emphasized: "Books seem to play a dual role in proving … information: On the one hand, they are effective in providing general information needed to formulate a scientific solution and on the other, specific information needed to choose a data analysis technique."

As is the case with this effort, preparation of any such compilation is thwarted with the problem that by the time it reaches the intended audience, it will be out of date. The generation of this collection of book titles was inspired to some degree by two such collections that are found in books authored by Roboz in 1968 (Reference 86) and Kiser in 1965 (Reference 90). Both of these collections were found to have been invaluable in my study of mass spectrometry. More recent collections of book titles in mass spectrometry appeared as part of the references in McLafferty and Tureček (Interpretation 7) and de Hoffman (Introductory 5). Both of these collections, like the Roboz and Kiser collections, are dated even though both were prepared in the 1990s.

This bibliography of mass spectrometry book titles of interest to mass spectrometrists is far from complete. Most of the entries are of books that have *mass spectrometry* in their titles. Only a few of the wide array of books about liquid chromatography, gas chromatography, and other forms of the separation sciences that go hand-in-glove with mass spectrometry are included. A good example of the wide range of books that mass spectrometry is part and parcel of is the nine books either authored by or edited by Jehuda Yinon, the noted Israeli expert on the analysis of explosive and other forensic areas. Only three of these books are included because they are the only three that contain *mass spectrometry* in their title. However, these other titles attributed to Yinon often include valuable information on mass spectral techniques.

The old adage "You can't judge a book by its cover!" is truer today than ever before. Although there has been a number of outstanding mass spectrometry books that address the new technologies developed at the end of the 20th century, the literature of mass spectrometry has been cursed with a plethora of recently published books that suffer from bad technical content, little or no competent copyediting, and/or amateurish and sloppy production. These problems are due to consolidations in the publishing industry resulting in fewer publishers, a perceived need for new books to replace older ones and to provide information on a subject that has exploded in the last decade with new ionization techniques, and the lack of desire and ability of authors to research their subject and avoid the self-gratification of generating an avalanche of neologisms and technical errors. These types of problems were practically nonexistent in books on mass spectrometry before 1990.

There is a well-written and organized general text on mass spectrometry (Introductory 11) that suffers from an incompetent production effort. In this particular book, there is a lack of consistency in the fonts used for the presentation of reaction schemes, figures are improperly imported from electronic submissions, and promised graphics on the inside cover are replaced with an easy-to-lose insert. This book is the singularly best modern text on mass spectrometry partially because its thousands of journal references all have titles. It is a shame that the publisher did not make the effort to do a better job of the presentation. Poor production efforts by established publishers is one of the reasons why you see more self-publishing (Introductory 3 and 7; Technique 15).

Another book, produced by a major publisher, consistently uses the word "spectra" (the plural form) for the word "spectrum" (the singular form) (Technique 38). As a corollary to you can't judge a book by its cover, "You can't necessarily judge a book by its title." A book was published with the title *Understanding Mass Spectra: A Basic Approach* (Technique 36) when, in reality, it should have been entitled *How I Came to Understand the Mass Spectra of Illicit Drugs: an Autobiographical Presentation*. There are also problems with poorly translated foreign-language books such as one originally written in French and translated into English by a person whose native language was Chinese (Introductory 5, 1st ed.); reviewed in *J. Am. Soc. Mass Spectrom.* **1997**, *8*, 1193, 1194. The second edition of this book, which appears to have been written in English, is far improved. Another problem with translated books is the delay in information. This dated information is especially a problem with the rapid change that is taking place in LC/MS and mass spectrometry and biotechnology.

Several books with copyright dates in the 1990s have been reviewed by me and others for the *Journal of the American Society for Mass Spectrometry*. When a review is available, it is cited along with the book. One of the books I reviewed is in the **Technique-Oriented Books** section (Technique 38: *J. Am. Soc. Mass Spectrom.* **1999**, *10*, 364–367) and one is in the **Interpretation Books** section (Interpretation 4: *J. Am. Soc. Mass Spectrom.* **1998**, *9*, 852–854). Both of these books were found to be of little or no value, and it was felt that they would be more harmful to the reader than helpful. Another book in the

Technique-Oriented Books section (Technique 2), by the same author as Technique 38, that has a similar title *LC/MS: A Practical User's Guide* is just as badly written and has just as much misinformation. What makes the publication of this book more egregious than the first book are the poor reviews of the first and the fact that when the book proposal was sent for peer review, the proposal was found to be amateurish and without merit. It was published because the publisher knew this title would sell. These three books should be avoided.

Most of the books on mass spectrometry come from the chemists who use the technique. Chemistry, unlike biochemistry and the biological sciences, has been slow to recognize the importance of the titles of journal articles in cited references. Many of the mass spectrometry books published in the last 4 years, most of them from the 10 years previous to that and almost all of them prior to the previous 10 years, do not include titles with cited journal articles. That is why a special effort has been made to mark books that do use titles of cited journal articles in this bibliography. These books are noted with an asterisk (*).

This bibliography is presented in 14 segments. Book-related segments include **Introductory**; **Reference**; **Technique-Oriented**; **Conference Proceedings**; **Interpretation**; **Ion/Molecule Reactions**; **Historical Significance**; **Collections of Mass Spectra in Hardcopy**; **Inductively Coupled Plasma Mass Spectrometry**; and **Integrated Spectral Interpretation**. Non-book-related segments include **Mass Spectrometry**, **GC/MS, and LC/MS Journals**; **Personal Computer MS Abstract Sources**; **Monographs**; and **Software**. These segments are included because of their importance in finding information on mass spectrometry.

There is no inclusion of specific articles from journals. Through the 1960s, such listings of journal articles were published in various forms—often by mass spectrometry instrumentation manufacturers. However, since the development of comprehensive abstract systems and their electronic availability, these printed listings no longer have much value. The **Personal Computer MS Abstract Sources** section is a guide to information on current journal articles. (See the note associated with item 24 in the **Historical Significance** segment for information on compilations of mass spectrometry journal articles.)

There is no inclusion of Internet search engines, such as Medline and Chemical Abstracts Service. Some of these search engines are found on the **Important World Wide Web URLs** page on the inside back cover of this book.

There has been some duplication of book titles. For example, the Watson book (Introductory 11 and Interpretation 5) covers two areas of mass spectrometry. This book is excellent for the interpretation of electron ionization spectra as well as an introductory book. All books are listed in chronological order by section. References to reviews that have appeared in the *Journal of the American Society for Mass Spectrometry* are noted.

Since the first edition of this book was published in January of 2000, not only have new bibliographical segments and recently published books been added, but other books that have older publication dates that were not in the first edition have also been added. The addition of segments has resulted in some rearrangements from the first edition. Whereas some segments may appear to contain fewer titles, previously listed titles have been moved to other segments. Another addition to this edition is the international standard book number (ISBN) where applicable and when available. The numbers are presented in a contiguous string.

Introductory Books

This section contains introductory books on mass spectrometry and gas and liquid chromatography. If you are going to own only one mass spectrometry book, it should be *Introduction to Mass Spectrometry*, 3rd edition (Introductory 11) (with a 4th edition soon to be published), which is a comprehensive book that touches on all the current technology. This book provides an excellent understanding of electron ionization (EI) fragmentation mechanisms (which is essential to the understanding of all fragmentation in mass spectrometry); and, of great importance, all of the journal articles referenced have titles. Another important book in this section is *Pushing Electrons: A Guide for Students of Organic Chemistry* (Introductory 9). This student workbook instills a good understanding of moving electrons within organic ions and molecules, which is also essential to the understanding of mass spectral fragmentation. The Watson book, along with *Pushing Electrons*, is a good choice for an introductory course. The books by Jürgen H. Gross and E. de Hoffmann are also good textbook choices, with preference to the Gross book.

MS Fundamentals (Introductory 14) is not a book. It is a multimedia training tool consisting of a computer-based training program, video, and book. This entry is also listed in the **Software** section. To those who work with the transmission quadrupole mass spectrometer, this multimedia presentation is indispensable. In order to get the most from any analytical instrument, you must have a thorough understanding of the technology. This multimedia presentation will result in that understanding, even in those who have little technical background.

There are three *Analytical Chemistry by Open Learning* books (Introductory 6, 15, and 21) listed in this section. These books are programmed-learning text for self-study; however, the mass spectrometry book is of questionable quality according to a number of reviewers and may be problematic as an introduction to the subject. Two of the entries in this section are video courses produced by the American Chemical Society (Introductory 20 and 24). Neither is recommended because the mass spectrometry course is dated, and the gas chromatography course has too little emphasis on capillary columns.

1. Downard, K *Mass Spectrometry: A Foundation Course*; The Royal Society of Chemistry: Cambridge, U.K., 2004, ISBN:0854046097.

2. *Gross, JH *Mass Spectrometry: A Textbook*; Springer-Verlag, Heidelberg, Germany, 2004, ISBN:3540407391 (reviewed *JASMS* 16:793).

3. Siuzdak, G *The Expanding Role of Mass Spectrometry in Biotechnology*; MCC: San Diego, CA, 2003, ISBN:0974245100.

4. Dreher, W *Moderne Massenspektrometric: Grundlagen, Kopplungs-Und Ionisationstechniken*; Wiley: Chichester, U.K., 2002, ISBN:3527303162.

5. de Hoffmann, E; Stroobant, V *Mass Spectrometry: Principles and Applications*, 2nd ed.; Wiley: New York, 2001, ISBN:0471485659 (original French language edition, *Spectrométrie de masse*, © Dunod, Paris, 1999); 1st ed., de Hoffmann, E; Charette, J; Stroobant, V *Mass Spectrometry: Principles and Applications*; © Masson éditeur Paris, 1996; Wiley: New York, 1996, ISBN:0471966967 (reviewed *JASMS* 8:1193) (original French language edition, *Spectrométrie de masse*, © Masson éditeur, Paris, 1994).

6. Barker, J *Mass Spectrometry: Analytical Chemistry by Open Learning*, 2nd ed.; Ando, DJ, Ed.; Wiley: Chichester, U.K., 1999, ISBN:0471967645 hc; ISBN:0471967629 pbk; 1st ed., Davis, R; Frearson, MJ, 1987, ISBN:047191388X hc; ISBN:0471913898 pbk (reviewed *JASMS* 4:831).

7. Cunico, RL; Gooding, KM; Wehr, T *Basic HPLC and CE of Biomolecules*; Bay Bioanalytical Laboratory: Richmond, CA, 1998, ISBN:0966322908.

8. Budzikiewicz, H *Massenspektrometrie – Eine Einführung,* 4th aufl. (*Mass Spectrometry: An Introduction*, 4th ed.); VCH: Weinheim, 1998.

9. Weeks, DP *Pushing Electrons: A Guide for Students of Organic Chemistry,* 3rd ed.; Saunders College: Fort Worth, TX, 1998, ISBN:0030206936.

10. McNair, HM; Miller, JM *Basic Gas Chromatography*; Wiley: New York, 1997, ISBN:047117260X hc; ISBN:0471172618 pbk.

11. *Watson, JT *Introduction to Mass Spectrometry*, 3rd ed.; Lippincott-Raven: Philadelphia-New York, 1997, ISBN:0397516886; 1st ed., *Introduction to Mass Spectrometry: Biomedical, Environmental, and Forensic Applications*; Raven: New York, 1976, ISBN:0890040567; 2nd ed., Raven: New York, 1985, ISBN:0881670812.

12. Siuzdak, G *Mass Spectrometry for Biotechnology*; Academic: San Diego, CA, 1996, ISBN:0126474710 (reviewed *JASMS* 7:1179).

13. Johnstone, RAW; Rose, ME *Mass Spectrometry for Chemists and Biochemists*, 2nd ed.; Cambridge University Press: Cambridge, U.K., 1996, ISBN:0521414660 hc; ISBN:0521424976 pbk (reviewed *JASMS* 9:649).

14. *MS Fundamentals: Multimedia Training*; Academy Savant: Fullerton, CA, 1995.

15. Fowlis, IA *Gas Chromatography: Analytical Chemistry by Open Learning*, 2nd ed.; Wiley: Chichester, U.K., 1995.

16. Russell, DH, Ed. *Experimental Mass Spectrometry*; Plenum: New York, 1994, ISBN:0306444577 (reviewed *JASMS* 6:277).

17. Hinshaw, JV; Ettre, LS *Introduction to Open-Tubular Column Gas Chromatography*; Advanstar: Cleveland, OH, 1993.

18. Ettre, LS; Hinshaw, JV *Basic Relationships of Gas Chromatography*; Advanstar: Cleveland, OH, 1993.

19. Chapman, JR *Practical Organic Mass Spectrometry*, 2nd ed., *A Guide for Chemical and Biochemical Analysis*; Wiley: New York, 1993, ISBN:0471927538; 1st ed., 1985, ISBN:0471906964.

20. Watson, JT *Introduction to Mass Spectrometry*; ACS Video Course; American Chemical Society: Washington, DC, 1993; 2 tapes, 130 pp (reviewed *JASMS* 7:298).

21. Lindsay, S *High Performance Liquid Chromatography: Analytical Chemistry by Open Learning,* 2nd ed.; Wiley: Chichester, U.K., 1992; 1st ed., 1987.

22. *Desiderio, DM, Ed. *Mass Spectrometry: Clinical and Biomedical Applications*, Vol. 1; Plenum: New York, 1992, ISBN:0306442612; Vol. 2, ISBN:0306444550 (reviewed *JASMS* 6:453).

23. Karasek, FW; Clement, RE *Basic Gas Chromatography-Mass Spectrometry*; Elsevier: New York, 1988.

24. McNair, HM *Basic Gas Chromatography*; ACS Video Course; American Chemical Society: Washington, DC, 1988.

25. Beynon, JH; Brenton, AG *Introduction to Mass Spectrometry*; University of Wales Press: Swansea, U.K., 1982, ISBN:0708308104.

26. Snyder, LR; Kirkland, JJ *Introduction to Modern Liquid Chromatography*, 2nd ed.; Wiley: New York, 1979; 1st ed., 1974 (Japanese translation, 1976).

27. Johnson, E; Stevenson, R *Basic Liquid Chromatography*; Varian: Palo Alto, CA, 1978.

28. Johnstone, RAW *Mass Spectrometry for Organic Chemist;* Cambridge University Press: Cambridge, U.K., 1972, ISBN:0521083818.

29. McNair, HM; Bonelli, EJ *Basic Gas Chromatography*, 5th ed.; Varian: Palo Alto, CA, 1969.

Reference Books

The citations in this section are meant to be general references on mass spectrometry. In addition, there is a reference to LC method development. One of the most important entries in this section is the "Mass Spectrometry" article in the biennial Fundamental Reviews issue of *Analytical Chemistry*. This review began in 1949 with the article authored by John A. Hipple and Martin Shepherd (National Bureau of Standards, currently the National Institute of Standards and Technology) with 165 citations. This review appeared in 1950 and 1952. Although there was a Fundamental Reviews issue in 1951, there was no mass spectrometry review. It was decided in 1953 that the Applications and Fundamental Reviews would appear in alternating years. Until that time, the Fundamental Reviews had been Issue 1 and the Applications Reviews had been Issue 2 of the volume year. 1953 was the Applications Reviews year. Issue 1 of 1954 was the Fundamental Reviews issue. For the next five years (1955–1959), the reviews were published as Part II of the April (No. 4) issue. Beginning with the 1960 volume year, the Reviews issues were given an issue number but were not included in the sequential page numbering of the volume.

The latest (and as it turns out, the last) publication of these Fundamental Reviews of Mass Spectrometry (Reference 17) to appear in *Analytical Chemistry* had 1551 citations divided into 9 categories: Overview (5), Scope (173), Innovative Techniques and Instrumentation (364), Isotope Ratio Mass Spectrometry (89), High-Power Laser in Mass Spectrometry (51), Dissociation by Low-Intensity Infrared Radiation (18), Polymers (61), Peptides and Proteins (624), and Oligonucleotides and Nucleic Acids (166). The "Mass Spectrometry" articles in the Fundamental Reviews issues of *Analytical Chemistry* have had Alma L. Burlingame as their primary author since 1972 with 14 consecutive articles. Unfortunately, no mass spectrometry review appeared in the 2000 Fundamental Reviews issue of *Analytical Chemistry*. Some three months later (September 15, 2000, issue) a review entitled "Environmental Mass Spectrometry" did appear in *Analytical Chemistry*. Reviews on environmental mass spectrometry have appeared in the Fundamental and Applications Reviews issues every year thereafter (2001–2004). In the 2002 Fundamental Reviews issue, there were also articles on ICP MS and Mass Spectrometry of Chemical Polymers. The Mass Spectrometry Fundamental Reviews, which existed from the fourth year after the *Analytical Edition of Industrial and Engineering News* changed its name to *Analytical Chemistry* through 1998, no longer exists. This was the last of the literary annual bibliographies on mass spectrometry. Today, that role is filled by *CA Selects Plus: Mass Spectrometry* or the bibliography that appears each month at the end of the *Journal of Mass Spectrometry*.

An important aspect of using a mass spectrometer is "becoming one" with the instrument. In order to accomplish this unison, you must have an in-depth working understanding of the instrument. The multimedia training program distributed by Academy Savant, *MS Fundamentals* (Introductory 14 and Software 26), is an excellent aid to gaining this understanding of the transmission quadrupole mass spectrometer. In addition, the Dawson book (Reference 73) is a seminal reference for the transmission quadrupole and the quadrupole ion trap. The March/Hughes book (Reference 44) is the seminal reference for the quadrupole ion trap, and the Cotter book (Reference 22) is a likewise reference for the time-of-flight mass spectrometer. Care must be taken with respect to the Cotter title because he has edited a book (Reference 30) that does not provide the detailed understanding of the TOF-MS. The same is true for the March/Hughes book because of a similar title edited by March and Todd (Reference 27). There is no good reference for the current technology of magnetic-sector mass spectrometers. The best material on understanding the fundamentals of a magnetic-sector and/or double-focusing instrument is the Roboz book (Reference 86). The best understanding of the workings of FTMS is found in an article, "Fourier Transform Ion Cyclotron Resonance Mass Spectrometry", authored by Alan Marshall et al. in *Encyclopedia of Analytical Chemistry*, Vol. 13; Sparkman, OD, Section Ed.; Meyers, RA, Ed.; Wiley: Chichester, U.K., 2000, pp 11694–11728.

Working with mass spectrometers often requires the derivatization of analytes to obtain the best results. This section has three good books that aid in this task (Reference 34, 64, and 87). In addition, there are two other books that do not primarily pertain to mass spectrometry techniques: *Basic Vacuum Practice* from Varian (Reference 45) and *The Mass Spec Handbook of Service* by Manura (Reference 33).

It should be noted that several books of the 1960s and early 1970s have been reprinted in the past few years (Reference 73, 86, 94, 95, and 97). Republication is not a new practice. Several books were reprinted by a publisher other than the original publisher several years after the first printing. ASMS plans to reprint several other books of historical and technological significance over the next few years.

1. *Caprioli, RM, Ed. *Biological Applications, Part B: Carbohydrates, Nucleic Acids and other Biological Compounds*, Vol. 3; Gross, ML; Caprioli, RM, Editors-in-Chief; Elsevier: Oxford, U.K., 2006, ISBN:0080438032.

2. *Todd, JF; March, RE *Quadrupole Ion Trap Mass Spectrometry*, 2nd ed.; Wiley–Interscience: New York, 2005, ISBN:0471488887 pbk; 456 pp.

3. Kaltashov, IA; Eyles, SJ *Mass Spectrometry in Biophysics: Conformation and Dynamics of Biomolecules*; Wiley: New York, 2005, ISBN:0471456020.

4. *Nibbering, NMM, Ed. *Fundamentals of and Applications to Organic (and Organometallic) Compounds* in *The Encyclopedia of Mass Spectrometry*, Vol. 4; Gross, ML; Caprioli, R, Editors-in-Chief; Elsevier: Oxford, U.K., 2005, ISBN:0080438466.

5. *Caprioli, RM; Gross, ML, Eds. *Biological Applications, Part A: Peptides and Proteins* in *The Encyclopedia of Mass Spectrometry*, Vol. 2; Gross, ML; Caprioli, R, Editors-in-Chief; Elsevier: Oxford, U.K., 2005, ISBN:0080438008.

6. Ashcroft, AE; Brenton, G; Monaghan, JJ, Eds. *Advances in Mass Spectrometry*, Vol. 16; Plenary and Keynote Lectures of the 16th International Mass Spectrometry Conference, Edinburgh, U.K., 31 August–5 September 2003; Elsevier: Amsterdam, 2004, ISBN:044451283.

7. Gauglitz, G; Vo-Dinh, T, Eds. *Handbook of Spectroscopy*, 2 volumes; Wiley-VCH Verlag GmbH: Weinheim, Germany, 2003, ISBN:3527297820.

8. Schalley, CA, Ed. *Modern Mass Spectrometry*, Vol. 225 *Topics in Current Chemistry*; Springer-Verlag: Berlin, Germany, 2003, ISBN:3540000984.

9. *Armentrout, PB, Ed. *Theory and Ion Chemistry* in *The Encyclopedia of Mass Spectrometry*, Vol. 1; Gross, ML; Caprioli, R, Eds.; Elsevier: Oxford, U.K., 2003, ISBN:0080438024.

10. Herbert, CG; Johnstone, RAW *Mass Spectrometry Basics*; CRC: Boca Raton, FL, 2003, ISBN:0849313546.

11. Emilo G, Ed. *Advances in Mass Spectrometry*, Proceedings of the 15th International Conference, 2000, Barcelona, Spain; Wiley: Chichester, U.K., 2001, ISBN:0471891533.

12. *Sparkman, OD *Mass Spectrometry* in *Encyclopedia of Analytical Chemistry*, Vol. 13; Sparkman, OD, Section Ed.; Meyers, RA, Ed.; Wiley: Chichester, U.K., 2000, ISBN:0471976709.

13. *Sparkman, OD *Mass Spectrometry Desk Reference*; Global View: Pittsburgh, PA, 2000, ISBN:0966081323 (reviewed *JASMS* 11:1144).

14. Lindon, JC; Tranter, GE; Holmes, JL, Eds. *Encyclopedia of Spectroscopy and Spectrometry*; Academic: San Diego, CA, 2000, ISBN:0122266803.

15. *Sparkman, OD (p 2604); Wells, GJ; Huston, CK (p 2662); Adams, F, et al. (p 2650). In *Encyclopedia of Environmental Analysis and Remediation*, Vol. 4; Meyers, RA, Ed.; Wiley: New York, 1998.

16. Karjalainen, EJ; Hesso, AE; Jalonen, JE; Karjalainen, UP, Eds. *Advances in Mass Spectrometry*, Proceedings of the 14th International Conference, 1997, Tampere, Finland; Elsevier: Amsterdam, 1998 (also available as a CD ROM).

17. Burlingame, AL; Boyd, RK; Gaskell, SJ "Mass Spectrometry" in the Fundamental Reviews issue of *Anal. Chem.* **1998**, *70*(16).

18. Tuniz, C; Tuniz, JR; Fink Bird, D, Eds. *Accelerator Mass Spectrometry: Ultrasensitive Analysis for Global Science*; CRC: Boca Raton, FL, 1998.

19. Grove, HE *From Hiroshima to the Iceman: The Development and Applications of Accelerator Mass Spectrometry*; American Institute of Physics: New York, 1998.

20. Platzner, LT; Habfast, K; Walder, A; Goetz, A *Modern Isotope Ratio Mass Spectrometry*; Wiley: New York, 1997.

21. Wilkins, CL, Ed. Mass Spectrometry. In *Handbook of Instrumental Techniques for Analytical Chemistry*; Settle, FA, Ed.; Prentice Hall: Upper Saddle River, NJ, 1997; Section V.

22. Cotter, RJ *Time-of-Flight Mass Spectrometry: Instrumentation and Applications in Biological Research*; American Chemical Society: Washington, DC, 1997, ISBN:0841234744 (reviewed *JASMS* 9:1104).

23. Snyder, LR; Kirkland, JJ; Glajch, JL *Practical HPLC Method Development*, 2nd ed.; Wiley: New York, 1997.

24. Ashcroft, AE *Ionization Methods in Organic Mass Spectrometry*, RSC Analytical Spectroscopy Monographs; Royal Society of Chemistry: Cambridge, U.K., 1997.

25. Baer, T; Ng, C-Y; Powis, I, Eds. *Large Ions: Their Vaporization, Detection and Structural Analysis*; Wiley: New York, 1996.

26. *Townshend, A, et al., Eds. *Encyclopedia of Analytical Science*, Vol. 5 *Liq–Mic*; Academic: San Diego, CA, 1995, ISBN:0122267052.

27. March, RE; Todd, JFJ, Eds. *Practical Aspects of Ion Trap Mass Spectrometry*; Vol. I, ISBN:0849344522; Vol. II, ISBN:084938253X; Vol. III, ISBN:0849382513; CRC: Boca Raton, FL, 1995 (reviewed *JASMS* 10:80, 3 volumes).

28. Ghosh, PK *International Series of Monographs on Physics, 90: Ion Traps*; Oxford Science: New York, 1995.

29. Corndies, I; Horváth, Gy; Vékey, K, Eds. *Advances in Mass Spectrometry*, Proceedings of the 13th International Conference, 1995, Budapest, Hungary; Elsevier: Amsterdam, 1995.

30. Cotter, RJ, Ed. *Time-of-Flight Mass Spectrometry*; ACS Symposium Series 549; American Chemical Society: Washington, DC, 1994, ISBN:0841227713 (reviewed *JASMS* 7:123).

31. Schlag, EW, Ed. *Time-of-Flight Mass Spectrometry and its Applications*; Elsevier: Amsterdam, 1994, ISBN:0444818758 (reviewed *JASMS* 5:949).

32. Matsuo, T; Caprioli, RM; Gross, ML; Seyama, Y, Eds. *Biological Mass Spectrometry: Present and Future*; Wiley: New York, 1994.

33. Manura, JJ; Baker, CW, Eds. *The Mass Spec Handbook of Service*, Vol. 2; Scientific Instrument Services: Ringoes, NJ, 1993.

34. Blau, K; Halket, J, Eds. *Handbook of Derivatives for Chromatography*, 2nd ed.; Wiley: New York, 1993, ISBN:047192699X; 1st ed., Blau, K; King, G, Eds.; Heyden: London, 1978, ISBN:0855012064.

35. Voress, L, Ed. *Instrumentation in Analytical Chemistry 1988–1991*; American Chemical Society: Washington, DC, 1992, ISBN:084122191X hc; ISBN:0841222029 pbk (reviewed *JASMS* 4:286).

36. Kistemaker, PG; Nibbering, NM, Eds. *Advances in Mass Spectrometry*, Proceedings of the 12th International Conference, Vol. 12, Amsterdam, 26–30 August 1991; Elsevier: Amsterdam, 1992 (reviewed *JASMS* 5:53).

37. Clement, RE; Siu, KWM; Hill, Jr., HH *Instrumentation for Trace Organic Monitoring*; Lewis: Boca Raton, FL, 1991, ISBN:0873712137 (reviewed *JASMS* 4:755).

38. Standing, KG; Ens, W, Eds. *Methods and Mechanisms for Producing Ions from Large Molecules*; Plenum: New York, 1991, ISBN:0306440172 (reviewed *JASMS* 3:780).

39. Asmoto, B, Ed. *FT-ICR/MS: Analytical Applications of Fourier Transform Ion Cyclotron Resonance Mass Spectrometry*; Wiley: New York, 1991.

40. Marshall, AG; Verdun, FR *Fourier Transforms in NMR, Optical and Mass Spectrometry: A User's Handbook*; Elsevier: Amsterdam, 1990.

41. Constantin, E; Schnell, A (Chalmers, MH, Translator) *Mass Spectrometry*; Ellis Horwood: Chichester, U.K., 1990, ISBN:0135555256 hc; ISBN:0135533635 pbk (reviewed *JASMS* 4:82) (original French language edition, *Spectrométrie de masse*; Tec & Doc, France, © the copyright holders).

42. Lubman, DM, Ed. *Lasers in Mass Spectrometry*; Oxford University Press: Oxford, U.K., 1990.

43. Meuzelaar, HLC; Isenhour, TL, Eds. *Computer-Enhanced Analytical Spectroscopy*, Vol. 2; Plenum: New York, 1990.

44. *March, RE; Hughes, RJ *Quadrupole Storage Mass Spectrometry*; Wiley: New York, 1989.

45. Varian Vacuum Products Division *Basic Vacuum Practice*, 2nd ed.; Varian: Palo Alto, CA, 1989.

46. Benninghoven, A, Ed. *Ion Formation from Organic Solids: Mass Spectrometry of Involatile Materials*; Wiley: New York, 1989.

47. Heinrich, N; Schwarz, H. In *Ion and Cluster Ion Spectroscopy and Structure*; Maier, JP, Ed.; Elsevier: Amsterdam, 1989.

48. Prokai, L *Field Desorption Mass Spectrometry*; Marcel Dekker: New York, 1989.

49. Middleditch, BS *Analytical Artifacts: GC, MS, HPLC, TLC, and PC*, Journal of Chromatography Library, Vol. 44; Elsevier: Amsterdam, 1989.

50. Hugli, TE, Ed. *Techniques in Protein Chemistry*; Academic: Orlando, FL, 1989, ISBN:0126820007 hc; ISBN:0126820015 pbk (reviewed *JASMS* 4:519).

51. Buchanan, MV, Eds. *Fourier Transform Mass Spectrometry*; American Chemical Society: Washington, DC, 1987.

52. Gray, NAB *Computer-Assisted Structure Elucidation*; Wiley: New York, 1986.

53. Futrell, JH, Ed. *Gaseous Ion Chemistry and Mass Spectrometry*; Wiley: New York, 1986.

54. Duckworth, HE; Barber, RC; Venkalasubramanian, VS *Mass Spectroscopy*, 2nd ed.; Cambridge University Press: Cambridge, U.K., 1986; 1st ed., 1958.

55. White, FA; Wood, GM *Mass Spectrometry: Applications in Science and Engineering*; Wiley: New York, 1986.

56. Märk, TD; Dunn, GH *Electron Impact Ionization*; Springer-Verlag: Berlin, Germany, 1985, ISBN:3211817786 Wien; ISBN:0378817786 New York.

57. Beynon, JH; McGlashan, ML, Eds. *Current Topics in Mass Spectrometry and Chemical Kinetics*; Heyden: London, 1982.

58. Longevialle, Pierre *Principes De La Spectrométrie De Masse Des Substances Organiques*; Masson: Paris, 1981.

59. Howe, I; Williams, D; Bowen, RD *Mass Spectrometry: Principles and Applications*, 2nd ed.; McGraw-Hill: New York, 1981, ISBN:007030575 hc; ISBN:0070705690 pbk.

60. de Mayo, P, Ed. *Rearrangements in Ground and Excited States*, Vols. 1–3; Academic: New York, 1980.

61. Schlunegger, UP *Advanced Mass Spectrometry: Applications in Organic and Analytical Chemistry*; Crompton, TR, Translation Ed., Pergamon: Oxford, U.K., 1980, ISBN:0080238424.

62. Merritt, C, Jr.; McEwen, CN, Eds. *Practical Spectroscopy Series*, Vol. 3 *Mass Spectrometry: Part B*; Marcel Dekker: New York, 1980.

63. Merritt, C, Jr.; McEwen, CN, Eds. *Practical Spectroscopy Series*, Vol. 3 *Mass Spectrometry: Part A*; Marcel Dekker: New York, 1979.

64. Knapp, DR *Handbook of Analytical Derivatization Reactions*; Wiley–Interscience: New York, 1979.

65. Levsen, K *Fundamental Aspects of Organic Mass Spectrometry*; Verlag Chemie: Weinheim, Germany, 1978. Note: This book is Volume 4 of a series entitled *Progress in Mass Spectrometry Fortschritte der Massenspektrometrie* edited by Herausegeben von Herbert Budzikiewicz. Vol. 1: Hesse, M *Indolakaloide*, Teil 1 (Text), Teil 2 (Spektren); Vol. 2: Drewes, SE *Chroman and Related Compounds*; Vol. 3: Hesse, M; Bernhard, HO *Alkaloide (außer Indol-, Triterpen- und Steroidalkaloide)*. The publication dates of these three previous volumes are not known, nor is it known if there are subsequent volumes.

66. Millard, BJ *Quantitative Mass Spectrometry*; Heyden: London, 1978.

67. Cooks, RG, Ed. *Collision Spectroscopy*; Plenum: New York, 1978.

68. Hatman, H; Wanczek, K-P *Ion Cyclotron Resonance Spectrometry – Lecture Notes in Chemistry*; Berthier, G, et al., Eds.; Springer-Verlag: Berlin, 1978, ISBN:0387087605.

69. Majer, JR *The Mass Spectrometer*; Taylor and Francis: Bristol, PA, 1977.

70. Gudzinowicz, BJ; Gudzinowicz, MJ; Martin, HF *Fundamentals of Integrated GC-MS* (in three parts), *Part III: The Integrated GC-MS Analytical System*; Marcel Dekker: New York, 1977.

71. Gudzinowicz, BJ; Gudzinowicz, MJ; Martin, HF *Fundamentals of Integrated GC-MS* (in three parts), *Part II: Mass Spectrometry*; Marcel Dekker: New York, 1976.

72. Gudzinowicz, BJ; Gudzinowicz, MJ; Martin, HF *Fundamentals of Integrated GC-MS* (in three parts), *Part I: Gas Chromatography*; Marcel Dekker: New York, 1976.

73. Dawson, PH, Ed. *Quadrupole Mass Spectrometry and Its Applications*; Elsevier: Amsterdam, 1976, ISBN:0444413456 (reprinted by the American Institute of Physics: Woodbury, NY, 1995).

74. Lehman, TA; Bursey, MM *Ion Cyclotron Resonance Spectrometry*; Wiley: New York, 1976, ISBN:047112530X.

75. Frigerio, A *Essential Aspects of Mass Spectrometry*; Spectrum: Flushing, NY, 1974, ISBN:0470281200.

76. Neeter, R; Kort, CWF *Metastable Precursor Ions. A Table for Use in Mass Spectrometry*; Elsevier: Amsterdam, 1973.

77. Waller, GR, Ed. *Biochemical Applications of Mass Spectrometry*; Wiley–Interscience: New York, 1972; Waller, GR; Dermer, OC, 1st supplement, 1980.

78. Williams, DH; Howe, I *Principles of Organic Mass Spectrometry*; McGraw-Hill: London, 1972.

79. Maccoll, A, Ed. *Mass Spectrometry*; MTP Int. Rev. Sci.; Physical Chemistry Series One, Vol. 5; Buckingham, AD, Consulting Ed.; Butterworths: London, 1972.

80. Meisel, WS *Computer Orientated Approaches to Pattern Recognition*; Academic: New York, 1972.

81. Milne, GWA, Ed. *Mass Spectrometry: Techniques and Applications*; Wiley–Interscience: New York, 1971.

82. Williams, DH, Ed. *Mass Spectrometry*, Vol. 1, 1971 and Vol. 2, 1973; Chemical Society: London.

83. Field, FH; Franklin, JL *Electron Impact Phenomena and the Properties of Gaseous Ions*, revised edition; Academic: New York, 1970; 1st ed., 1957.

84. Melton, CE *Principles of Mass Spectrometry and Negative Ions*; Marcel Dekker: New York, 1970.

85. Kientiz, H; Aulinger, FG; Habfast, K; Spiteller, G *Mass Spectrometry*; Verlag Chemie: Weinheim, Germany, 1968.

86. Roboz, J *Introduction to Mass Spectrometry Instrumentation and Techniques*; Wiley: New York, 1968 (reprinted by ASMS, 2000).

87. Pierce, AE *Silylation of Organic Compounds*; Pierce Chemical: Rockford, IL, 1968.

88. White, FA *Mass Spectrometry in Science and Technology*; Wiley: New York, 1968.

89. Tatematsu, A; Tsuchiya T, Editors-in-Chief *Structure Indexed Literature of Organic Mass Spectra, 1966*; Organic Mass Spectral Data Division, Society of Mass Spectrometry of Japan; Academic Press of Japan: Tokyo, 1968.

90. Kiser, RW *Introduction to Mass Spectrometry and Its Application*; Prentice-Hall: Englewood Cliffs, NJ, 1965.

91. Beynon, JH; Sanders, RA; Williams, AE *Table of Meta-stable Transitions*; Elsevier: New York, 1965.

92. Lederberg, J *Computation of Molecular Formulas for Mass Spectrometry*; Holden-Day: San Francisco, CA, 1964.

93. Beynon, JH; Williams, AE *Mass and Abundance Tables for Use in Mass Spectrometry*; Elsevier: New York, 1963.

94. McDowell, CA, Ed. *Mass Spectrometry*; McGraw-Hill: New York, 1963 (reprinted by Robert E. Krieger: Huntington, NY, 1979).

95. Biemann, K *Mass Spectrometry: Organic Chemical Applications*; McGraw-Hill: New York, 1962 (reprinted by ASMS, 1998).

96. Reed, RI *Ion Production by Electron Impact*; Academic: London, 1962.

97. Beynon, JH *Mass Spectrometry and Its Applications to Organic Chemistry*; Elsevier: Amsterdam, 1960 (reprinted by ASMS, 1999).

98. Barnard, GP *Modern Mass Spectrometry*; American Institute of Physics: London, 1953.

Technique-Oriented Books

The books listed in this section pertain to specific techniques of mass spectrometry and hyphenated chromatography/mass spectrometry techniques. In some cases, the books are specific to certain types of analytes. Books of this type began to appear at the end of the 1960s (only two listings before 1970, Technique 138 and 139). Slightly less than two/thirds of these books (87 out of 139) have copyright dates in the last decade and a half. Unfortunately, the last seven or eight years have seen a proliferation of very poorly written books with little or no copyediting and poor quality production.

Many of the technique-oriented books are made obsolete by changing technology within a few years or months of their publication. A good example is the 1990 Yergey book (Technique 79) on LC/MS. Because of the massive advancements in technology that have taken place in the last 15 years, this book had little relevance to the technique within 2 years of its publication. The book still has a great deal of value in that it provides good information on how to perform the chromatographic separations required in LC/MS, and it has an excellent set of journal-article references that contains titles. The problem of dated material can be especially significant with foreign-language books that are translated into English. Unfortunately, in some cases, publishers are not indicating that a book has been translated. It is only through careful research that the foreign-language roots of a book can be established, such as the case with Technique 34.

There are two books (Technique 15 and 52) that are of particular value to the LC/MS and GC/MS practitioner, respectively. These two books have a great deal of practical information on the running of different types of analyses and are good aids in the decision-making process about how to proceed with a particular sample. The Willoughby book (Technique 15) has very useful information in deciding whether to use a contract laboratory or perform the analysis in-house. The Willoughby book was reviewed in *JASMS* (*J. Am. Soc. Mass Spectrom.* **1999**, *10*, 78, 79) as was the Kitson book (*J. Am. Soc. Mass Spectrom.* **1997**, *9*, 294, 295). Care must be taken with respect to the unfortunate similarity between the title of the Kitson book (Technique 52) and the book title of Technique 38.

Books that are edited works rather than having a single author generally don't get my approval. These edited editions often end up looking like a "camel" (a horse designed by a committee). This lack of continuity in edited books is more true of books published in the last two and a half decades than those published before that time. There is one notable exception in Technique 45 by Cole (reviewed *J. Am. Soc. Mass Spectrom.* **1997**, *8*, 1191, 1192). This book is an excellent reference for those working in electrospray. The second edition of the Niessen book (Technique 39) is also a good reference for electrospray as well as other LC/MS techniques. The single negative about both of these books is that they do not include titles with journal-article citations.

There are two important references on environmental GC/MS that should be reviewed by anyone working in this area (Technique 118 and 120). Although both of these books were written in the era of the packed column, the fundamentals of environmental analyses and the U.S. Environmental Protection Agency (EPA) tune-criteria are covered in detail.

Another reference of the packed-column era is the McFadden book (Technique 132). This book, along with the Karasek book (Introductory 23), is very useful to those starting in GC/MS.

If you are using chemical ionization (either atmospheric pressure chemical ionization or chemical ionization under the conditions normally encountered in GC/MS), you need the Harrison book (Technique 66). Unlike the Yinon book (Technique 59), this book is a second edition and is labeled as such.

In looking at new titles of technique-orientated books, care must be taken to know when the book is nothing more than a collection of a series of articles from a journal or a bound issue of a journal. The value of such books is often less than their extremely high selling price.

1. Wilkins, CL; Lay, JO; Winefordner, JD, Eds. *Identification of Microorganisms by Mass Spectrometry*; Wiley: New York, 2005, ISBN:0471654426.

2. McMaster, MC *LC/MS: A Practical User's Guide*; Wiley: New York, 2005, ISBN:0471655317.

3. Lee, MS, Ed. *Integrated Strategies for Drug Discovery Using Mass Spectrometry*; Wiley: New York, 2005, ISBN:047146127X.

4. Henderson, W; McIndoe, JS *Mass Spectrometry in Biophysics: Conformation and Dynamics of Biomolecules*; Wiley: New York, 2005, ISBN:0470850159 hc; ISBN:0470850167 pbk.

5. Bienvenut, WV, Ed. *Acceleration and Improvement of Protein Identification by Mass Spectrometry*; Springer-Verlag: New York, 2005, ISBN:1402033184 hc; 298 pp.

6. Matthiesen, Ed. *Mass Spectrometry Data Analysis in Proteomics*; Humana: Totowa, NJ, 2005, ISBN:158829563X hc.

7. Alba, ARF *Chromatographic-Mass Spectrometric Food Analysis for Trace Determination of Pesticide Residues*, Vol. 73; Elsevier: Amsterdam, 2004, ISBN:0444509437 hc; 476 pp.

8. Korfmacher, WA, Ed. *Using Mass Spectrometry for Drug Metabolism Studies*; CRC: Boca Raton, FL, 2004, ISBN:0849319633.

9. *Yinon, J, Ed. *Advances in Forensics Applications of Mass Spectrometry*; CRC: Boca Raton, FL, 2004, ISBN:0849315220.

10. Ferrer, I; Thurman, EM *Liquid Chromatography/Mass Spectrometry MS/MS and Time-of-Flight MS: Analysis of Emerging Contaminants*; ACS Symposium Series 850; ACS: Washington, DC, 2003, ISBN:0841238251 (distributed by Oxford: New York).

11. Ardrey, RE *Liquid Chromatography-Mass Spectrometry: An Introduction*; Wiley: Chichester, U.K., 2003, ISBN:0471497991.

12. Pasch, H; Schrepp, W *MALDI-TOF Mass Spectrometry of Synthetic Polymers*; Springer Laboratory Series; Springer-Verlag: New York, 2003, ISBN:3540442596.

13. Simpson, RJ *Proteins and Proteomics: A Laboratory Manual*; Cold Spring Harbor Laboratory Press: Cold Spring Harbor, New York, 2003, ISBN:0879695536 hc; ISBN:0879695544 pbk.

14. Murphy, RC *Mass Spectrometry of Phospholipids: Tables of Molecular and Product Ions*; Illuminati: Denver, CO, 2002, ISBN:0970283415 spiral bound; 71 pp.

15. *Willoughby, R; Sheehan, E; Mitrovich, S *A Global View of LC/MS,* 2nd ed; Global View: Pittsburgh, PA, 2002, ISBN:0966081358; 1st ed., 1998, ISBN:0966081307 (reviewed *JASMS* 10:78).

16. Silberring, J; Ekman, R *Mass Spectrometry and Hyphenated Techniques in Neuropeptide Research*; Wiley: New York, 2002, ISBN:0471354937 (reviewed *JASMS* 14:81).

17. Pramanik, B; Ganguly, A; Gross, ML, Eds. *Applied Electrospray Mass Spectrometry*; Marcel Dekker: New York, 2002, ISBN:0824706188 (reviewed *JASMS* 13:1013, 1014).

18. Lee, MS *LC/MS Applications in Drug Development*; Wiley: New York, 2002, ISBN:0471405205.

19. Liebler, DC *Introduction to Proteomics*; Humana: Totowa, NJ, 2002, ISBN:0896039919 hc; ISBN:0896039927 pbk (reviewed *JASMS* 13:1148).

20. Sargent, M; Harrington, C; Harte, R, Eds. *Guidelines for Achieving High Accuracy in Isotope Dilution Mass Spectrometry (IDMS)*; RSC: Cambridge, U.K., 2002, ISBN:0854044183.

21. *Rossi, DT; Sinz, MW, Eds. *Mass Spectrometry in Drug Discovery*; Marcel Dekker: New York, 2001, ISBN:0824706072 (reviewed *JASMS* 13:1356, 1357).

22. Montaudo, G; Lattime, RP *Mass Spectrometry of Polymers*; CRC: Boca Raton, FL, 2001, ISBN:084930167X (reviewed *JASMS* 13:1250, 1251).

23. Roboz, J *Mass Spectrometry in Cancer Research*; CRC: Boca Raton, FL, 2001, ISBN:0849331277 (reviewed *JASMS* 14:81).

24. Dass, C *Principles and Practice of Biological Mass Spectrometry*; Wiley–Interscience Series on Mass Spectrometry; Desiderio, DM; Nibbering, NMM, Eds.; Wiley–Interscience: New York, 2001, ISBN:0471330531.

25. Budde, WL *Analytical Spectrometry Strategies for Environmental and Related Applications*; Oxford University Press: New York, 2001 (reviewed *JASMS* 13:898–900).

26. James, P, Ed. *Proteome Research: Mass Spectrometry*; Springer-Verlag: Berlin, 2001, ISBN:3540672559 hc; ISBN:3540672567 pbk.

27. Niessen, WMA, Ed. *Current Practice of Gas Chromatography–Mass Spectrometry*; Marcel Dekker: New York, 2001, ISBN:0824704738 (reviewed *JASMS* 12:1348).

28. Housby, JN *Mass Spectrometry and Genomic Analysis*; Kluwer Academic: Boston, MA, 2001, ISBN:0792371739.

29. Hübschmann, H-J *Handbook of GC/MS: Fundamentals and Applications*; Wiley-VCH: Weinheim, 2001, ISBN:3527301704.

30. Swartz, ME *Analytical Techniques in Combinatorial Chemistry*; Marcel Dekker: New York, 2000, ISBN:0824719395.

31. Mellon, F; Self, R; Startin, JR *Mass Spectrometry of Natural Substances in Food*; Belton, PS, Series Ed.; RSC: Cambridge, U.K., 2000, ISBN:0854045716.

32. *Kinter, M; Sherman, N *Protein Identification and Sequencing Using Tandem Mass Spectrometry*; Wiley–Interscience Series on Mass Spectrometry; Desiderio, DM; Nibbering, NMM, Eds.; Wiley–Interscience: New York, 2000, ISBN:04713224907.

33. *Chapman, JR *Mass Spectrometry of Proteins and Peptides*; Humana: Totowa, NJ, 2000.

34. Håkansson, K *Method and Technique Development in Peptide and Protein Mass Spectometry*; Comprehensive Summaries of Uppsala Dissertations from the Faculty of Science and Technology 518; Acta Universitatis Upsaliensis: Uppsala, Sweden, 2000, ISBN:9155446817 hc; ISBN:1104232X pbk.

35. Gerhards, P; Bons, U; Sawazki, J; Szigan, J; Wertmann, A *GC/MS in Clinical Chemistry*; Wiley: Chichester, U.K., 1999, ISBN:3527296239 (reviewed *JASMS* 12:125).

36. Smith, RM *Understanding Mass Spectra: A Basic Approach*; Busch, KL, Tech. Ed.; Wiley: New York, 1999, ISBN:0471297046 (reviewed *JASMS* 11:664).

37. Niessen, WMA *Liquid Chromatography-Mass Spectrometry*, 2nd ed.; Chromatographic Science Series, Vol. 79; Marcel Dekker: New York, 1999.

38. McMaster, M; McMaster, C *GC/MS: A Practical User's Guide*; Wiley: New York, 1998, ISBN:0471248266 (reviewed *JASMS* 10:364).

39. Niessen, WMA; Voyksner, RD, Eds. *Current Practice in Liquid Chromatography-Mass Spectrometry*; Elsevier: Amsterdam, 1998 (reprinted from *Journal of Chromatography A*, Vol. 794).

40. *Larsen, BS; McEwen, CN, Eds. *Mass Spectrometry of Biological Materials*; Marcel Dekker: New York, 1998.

41. *Thurman, EM; Mills, MS *Solid Phase Extraction: Principles and Practice*; Wiley: New York, 1998.

42. Meyers, RA, Ed. *Encyclopedia of Environmental Analysis and Remediation*; Wiley: New York, 1998.

43. Briggs, D; Ward, IM; Suresh, S; Clarke, DR, Eds. *Surface Analysis of Polymers by XPS and Static SIMS*; Cambridge University Press: Cambridge, U.K., 1998.

44. Oehme, M *Practical Introduction to GC-MS Analysis with Quadrupoles*; Huthig: Heidelburg, Germany; Wiley: New York, 1998.

45. Cole, RB, Ed. *Electrospray Ionization Mass Spectrometry: Fundamentals, Instrumentation, and Applications*; Wiley: New York, 1997, ISBN:0471145645 (reviewed *JASMS* 8:1191).

46. Platzner, IT *Modern Isotope Ratio Mass Spectrometry*; with contributions by Habfast, K; Walder, AJ; Goetz, A; Wiley: New York, 1997.

47. Newton, RP; Walton, TJ, Eds. *Proceedings of the Phytochemical Society of Europe, 40: Applications of Modern Mass Spectrometry in Plant Science Research*; Clarendon: Oxford, U.K., 1997.

48. Briggs, D; Ward, IM; Suresh, SJ; Clark, DR, Eds. *Surface Analysis of Polymers by XPS and Static SIMS*; Cambridge: London, 1997, ISBN:0521352223.

49. Hancock, WS *New Methods in Peptide Mapping for the Characterization of Proteins*; CRC: Boca Raton, FL, 1996.

50. Barcelo, D, Ed. *Applications of LC-MS in Environmental Chemistry*, Journal of Chromatography Library, Vol. 59; Elsevier: Amsterdam, 1996.

51. *Chapman, JR, Ed. *Protein and Peptide Analysis by Mass Spectrometry*; Humana: Totowa, NJ, 1996, ISBN:0896033457 (reviewed *JASMS* 8:296).

52. *Kitson, FG; Larsen, BS; McEwen, CN *Gas Chromatography and Mass Spectrometry: A Practical Guide*; Academic: San Diego, CA, 1996, ISBN:0124833853 (reviewed *JASMS* 8:294).

53. Mellon, F; Sandström, B, Eds. *Stable Isotopes in Human Nutrition: Inorganic Nutrient Metabolism*; Academic: San Diego, CA, 1996, ISBN:0124905404.

54. Karger, BL; Hancock, WS, Eds. *Methods in Enzymology*, Vols. 270 and 271 *High Resolution Separation and Analysis of Biological Macromolecules, Part A: Fundamentals* and *Part B: Applications*; Academic: San Diego, CA, 1996 (two separate books).

55. *Boutton, TW; Yamasaki, S-i, Eds. *Mass Spectrometry of Soils*; Marcel Dekker: New York, 1996, ISBN:0824796993 (reviewed *JASMS* 8:200).

56. *Walker, JM, Ed. *The Protein Protocols Handbook*; Humana: Totowa, NJ, 1996.

57. Wampler, TP, Ed. *Applied Pyrolysis Handbook*; Marcel Dekker: New York, 1995, ISBN:082479446X.

58. Snyder, AP, Ed. *Biochemical and Biotechnology Applications of Electrospray Ionization Mass Spectrometry*; ACS Symposium Series 619; American Chemical Society: Washington, DC, 1995.

59. Yinon, J, Ed. *Forensic Applications of Mass Spectrometry*; CRC: Boca Raton, FL, 1995.

60. Fenselau, C, Ed. *Mass Spectrometry for the Characterization of Microorganisms*; ACS Symposium Series 549; American Chemical Society: Washington, DC, 1994, ISBN:0841227373 (reviewed *JASMS* 6:1262).

61. Ardrey, B, Ed. *Liquid Chromatography/Mass Spectrometry*; VCH: New York, 1993.

62. *Murphy, RC *Handbook of Lipid Research*, No. 7 *Mass Spectrometry of Lipids*; Plenum: New York, 1993, ISBN:0306443619 (reviewed *JASMS* 5:124).

63. Vertes, A; Gijbels, R; Adams, F, Eds. *Laser Ionization Mass Analysis*, Chemical Analysis: A Series of Monographs on Analytical Chemistry and Its Applications, Vol. 124; Winefordner, JD, Ed.; Kolthoff, IM, Ed. Emeritus; Wiley–Interscience: New York, 1993, ISBN:0471536733.

64. Niessen, WMA; van der Greef, J *Liquid Chromatography-Mass Spectrometry*; Chromatographic Science Series, Vol. 58; Marcel Dekker: New York, 1992, ISBN:0824786351 (reviewed *JASMS* 6:222).

65. Gross, ML, Ed. *Mass Spectrometry in the Biological Sciences: A Tutorial*; NATO ASI Series C: Mathematical and Physical Sciences, Vol. 353; Kluwer Academic: Boston, MA, 1992, ISBN:0792315391 (reviewed *JASMS* 3:867).

66. *Harrison, AG *Chemical Ionization Mass Spectrometry*, 2nd ed.; CRC: Boca Raton, FL, 1992, ISBN:0849342546; 1st ed., 1983, ISBN:0849356164 (reviewed *JASMS* 4:286).

67. Ho, MH, Ed. *Analytical Methods in Forensic Chemistry*; Ellis Horwood: Chichester, U.K., 1992 (reviewed *JASMS* 4:362).

68. Angeletti, RH, Ed. *Techniques in Protein Chemistry III*; Academic: Orlando, FL, 1992, ISBN:0120587556 hc; ISBN:0120587564 pbk (reviewed *JASMS* 4:519).

69. Juinno, K *Hyphenated Techniques in Supercritical Fluid Chromatography and Extraction*; Elsevier: Amsterdam, 1992.

70. St. Pyrek, J Mass Spectrometry in the Chemistry of Natural Products. In *Recent Advances in Phytochemistry*, Vol. 25 *Modern Phytochemical Methods*; Fischer, NH, et al., Eds.; Plenum: New York, 1991; Chapter 6.

71. Czanderna, AW; Hercules, DM *Ion Spectroscopies for Surface Analysis*; Plenum: New York, 1991.

72. Czanderna, AZ; Hercules, DM *Spectroscopies for Surface Analysis*; Kluwer: Boston, MA, 1991, ISBN:0306437929.

73. Ng, CY, Ed. *Vacuum Ultraviolet Photoionization and Photodissociation of Molecules and Clusters*; World Scientific: Teaneck, NJ, 1991, ISBN:9810204302 hc; ISBN:9810204310 pbk (reviewed *JASMS* 4:974).

74. Villafranca, JJ, Ed. *Techniques in Protein Chemistry II*; Academic: Orlando, FL, 1991, ISBN:023732958 hc; ISBN:0127219579 pbk (reviewed *JASMS* 4:519).

75. Standing KG; Ens, K *Methods and Mechanisms for Producing Ions from Large Molecules*; Plenum: New York, 1991.

76. Vickerman, JC; Brown, AW; Reed, NM, Eds. *Secondary Ion Mass Spectrometry: Principles and Applications*; Oxford: Cambridge, U.K., 1990, ISBN:019855625X.

77. Hilf, ER, Ed. *Mass Spectrometry of Large Non-Volatile Molecules for Marine Organic Chemistry*; World Scientific: River Edge, NJ, 1990.

78. *Fox, A; Morgan, SL; Larsson, L; Odham, G, Eds. *Analytical Microbiology Methods: Chromatography and Mass Spectrometry*; Plenum: New York, 1990.

79. *Yergey, AL; Edmonds, CG; Lewis, IAS; Vestal, ML *Liquid Chromatography/Mass Spectrometry: Techniques and Applications*; Plenum: New York, 1990.

80. Halket, JM; Rose, ME *Introduction to Bench-Top GC/MS*; HD Science: Stapleford, U.K., 1990.

81. Suelter, CH; Watson, JT, Eds. *Methods of Biochemical Analysis*, Vol. 34 *Biomedical Applications in Mass Spectrometry*; Wiley–Interscience: New York, 1990.

82. McCloskey, JA, Ed. *Methods in Enzymology*, Vol. 193 *Mass Spectrometry*; Academic: San Diego, CA, 1990, ISBN:0121820947 (reviewed *JASMS* 3:779).

83. McEwen, CN; Larsen, BS, Eds. *Practical Spectroscopy Series: Mass Spectrometry of Biological Materials*; Marcel Dekker: New York, 1990.

84. Brown, MA, Ed. *Liquid Chromatography/Mass Spectrometry: Applications in Agricultural, Pharmaceutical and Environmental Chemistry*; ACS Symposium Series 420; American Chemical Society: Washington, DC, 1990, ISBN:0841217408 (reviewed *JASMS* 4:831).

85. Caprioli, RM, Ed. *Continuous-Flow Fast Atom Bombardment Mass Spectrometry*; Wiley: New York, 1990, ISBN:0471928631 (reviewed *JASMS* 3:867).

86. *Desiderio, DM, Ed. *Mass Spectrometry of Peptides*; CRC: Boca Raton, FL, 1990, ISBN:0849362938 (reviewed *JASMS* 4:519).

87. SCIEX, *The API Book*; SCIEX, Division of MDS Health Group: Mississauga, Ontario, Canada, 1990.

88. Dolan, JW; Snyder, LR *Troubleshooting LC Systems*; Humana: Totowa, NJ, 1989.

89. Wilson, RG; Stevie, FA; Magee, CW *Secondary Ion Mass Spectrometry: A Practical Handbook for Depth Profiling and Bulk Impurity Analysis*; Wiley: New York, 1989, ISBN:0471519456.

90. Ashe, TR; Wood, KV, Eds. *Novel Techniques in Fossil Fuel Mass Spectrometry*; ASTM: Washington, DC, 1989.

91. Lawson, AM, Ed. *Clinical Biochemistry: Principles, Methods, Applications,* Vol. 1 *Mass Spectrometry*; Curtius, H Ch; Roth, M, Series Eds.; Walter deGruyter: New York, 1989, ISBN:3110077515.

92. Biermann, CJ; McGinnis, GD, Eds. *Analysis of Carbohydrates by GLC and MS*; CRC: Boca Raton, FL, 1988, ISBN:0849368510 (reviewed *JASMS* 4:97).

93. Lai, S-TF *Gas Chromatography/Mass Spectrometry Operation*; Realistic Systems: East Longmeadow, MA, 1988.

94. Busch, KL; Glish, GL; McLuckey, SA, Eds. *Mass Spectrometry/Mass Spectrometry: Techniques and Applications of Tandem Mass Spectrometry*; VCH: New York, 1988.

95. Benninghoven, A; Werner, HW; Rudenauer, FG *Secondary Ion Mass Spectrometry: Basic Concepts, Instrumental Aspects, Applications and Trends*; Wiley: New York, 1987, ISBN:0471010561.

96. Yinon, J, Ed. *Forensic Mass Spectrometry*; CRC: Boca Raton, FL, 1987.

97. Gilbert, J, Ed. *Applications of Mass Spectrometry in Food Science*; Elsevier: London, 1987.

98. Heinzle, E; Reuss, M, Eds. *Mass Spectrometry in Biotechnological Process Analysis and Control*; Plenum: New York, 1987.

99. Rosen, JD, Ed. *Applications of New Mass Spectrometry, Techniques in Pesticide Chemistry*, Chemical Analysis: A Series of Monographs on Analytical Chemistry and Its Applications, Vol. 91; Winefordner, JD, Ed.; Kolthoff, IM, Editor Emeritus; Wiley–Interscience: New York, 1987.

100. Jaeger, H *Capillary Gas Chromatography Mass Spectrometry in Mass Spectrometry in Medicine and Pharmacology*; Huetig: New York, 1987.

101. *Linskens, HF; Jackson, JF, Eds. *Modern Methods of Plant Analysis: Gas Chromatography/Mass Spectrometry*; New Series, Vol. 3; Springer-Verlag: Berlin, 1986.

102. de Graeve, J; Berthou, F; Prost, M *Méthodes Chromatographiques Couplées à la Spectrométrie De Masse* with collaboration of Arpino, P and Promè, JC; Mason: Paris, 1986, ISBN:2225806276.

103. *Gaskell, SJ, Ed. *Mass Spectrometry in Biomedical Research*; Wiley: Chichester, U.K., 1986, ISBN:0471910457.

104. Aczel, T, Ed. *Mass Spectrometric Characterization of Shale Oils*; ASTM: Philadelphia, PA, 1986.

105. Lyon, PA, Ed. *Desorption Mass Spectrometry: Are SIMS and FAB the Same?*; ACS Symposium Series 291; American Chemical Society: Washington, DC, 1985.

106. Karasek, FW; Hutzinger, O; Safe, S, Eds. *Mass Spectrometry in Environmental Sciences*; Plenum: New York, 1985.

107. Facchetti, S, Ed. *Mass Spectrometry of Large Molecules*; Elsevier: Amsterdam, 1985.

108. Desiderio, DM *Analysis of Neuropeptides by Liquid Chromatography and Mass Spectrometry* in *Techniques and Instrumentation*; Analytical Chemistry Series, No. 6; Elsevier: New York, 1984, ISBN:0444424180.

109. Message, GM *Practical Aspects of Gas Chromatography/Mass Spectrometry*; Wiley: New York, 1984.

110. *Odham, G; Larsson, L; Mardh, P-A, Eds. *Gas Chromatography/Mass Spectrometry: Applications in Microbiology*; Plenum: New York, 1984, ISBN:0306413140.

111. Voorhees, KJ *Analytical Pyrolysis: Techniques and Applications*; Butterworth: London, 1984.

112. McLafferty, FW, Ed. *Tandem Mass Spectrometry*; Wiley–Interscience: New York, 1983.

113. Facchetti, S, Ed. *Applications of Mass Spectrometry to Trace Analysis*; Elsevier: Amsterdam, 1982.

114. *Goodman, SI; Markey, SP *Diagnosis of Organic Academias By Gas Chromatography-Mass Spectrometry*, Vol. 6 *Laboratory and Research Methods in Biology and Medicine*; Alan R. Liss: New York, 1981.

115. Meuzelaar, HLC; Haverkamp, J; Hileman, SD *Techniques and Instrumentation* in Analytical Chemistry Series, Vol. 3 *Pyrolysis Mass Spectrometry of Biomaterials*; Elsevier: Amsterdam, 1980, ISBN:0444420991.

116. Wolfe, RR *Radioactive and Stable Isotope Tracers in Biomedicine: Principles and Practice of Kinetic Analysis*; Wiley: New York, 1980, ISBN:0471561312 (reviewed *JASMS* 4:193).

117. Foltz, RL; Fentiman, AF, Jr.; Foltz, RB *GC/MS Assays for Abused Drugs in Body Fluids*, National Institute on Drug Abuse Research Monograph Series 32; Department of Health and Human Services, Public Health Service, National Institute on Drug Abuse: Rockville, MD, 1980, Superintendent of Documents, U.S. Government Printing Office: Washington, DC.

118. Budde, WL; Eichelberger, JW *Organics Analysis Using Gas Chromatography/Mass Spectrometry*; Ann Arbor Science: Ann Arbor, MI, 1979.

119. Middleditch, BS, Ed. *Practical Mass Spectrometry*; Plenum: New York, 1979.

120. Keith, LH, Ed. *Identification & Analysis of Organic Pollutants in Water*; Ann Arbor Science: Ann Arbor, MI, 1979.

121. Land, DG; Nursten, HE *Progress in Flavour Research*; Applied Science: London, 1979.

122. *Payne, JP; Bushman, JA; Hill, DH, Eds. *The Medical and Biological Applications of Mass Spectrometry*; Academic: London, 1979, ISBN:0125479506.

123. Tatematsu, A; Miyazaki, H; Suzuki, M; Maruyama, Y *Practical Mass Spectrometry for the Medical and Pharmaceutical Sciences*; Kodansha: Tokyo, 1979 (a translation of Igaku to Yakugaku no Tame no Masu Supekutorometori, 1975).

124. Gudzinowicz, BJ; Gudzinowicz, MJ, Eds. *Analysis of Drugs and Metabolites by Gas Chromatography Mass Spectrometry*, Vol. 1 *Respiratory Gases, Ethyl Alcohol, and Related Toxicological Materials*, 1977; Vol. 2 *Hypnotics, Anticonvulsants, and Sedatives;* Vol. 3 *Antipsychotics, Antiemetics, and Antidepressant Drugs*; Vol. 4 *Central Nervous System Stimulants*; Vol. 5 *Analgesics, Local Anaesthetics, and Antibiotics*, 1978; Vol. 6 *Cardiovascular, Antihypertensive, Hypoglycemic, and Tiered-Related Agents*, 1979; Vol. 7 *Natural, Pyrolytic, and Metabolic Products of Tobacco and Marijuana*, 1980; Marcel Dekker: New York.

125. Beckey, HD *Principles of Field Ionization and Field Desorption Mass Spectrometry*; Pergamon: New York, 1977, ISBN:0080206123.

126. Masada, Y *Analysis of Essential Oils by Gas Chromatography and Mass Spectrometry*; Halsted (Division of Wiley): New York, 1976 (© 1976, Hirokawa: Japan).

127. Zaretskii, ZV *Mass Spectrometry of Steroids*; Wiley: New York, 1976.

128. Leclercq, PA *Some Applications of Mass Spectrometry in Biochemistry*; Self-Published: The Netherlands, 1975.

129. Haque, R; Biros, FJ, Eds. *Mass Spectrometry and NMR Spectroscopy in Pesticide Chemistry*; Plenum: New York, 1974.

130. Melville, RS; Dobson, VF, Eds. *Selected Approaches to Gas Chromatography-Mass Spectrometry in Laboratory Medicine*, a one-day conference sponsored by the Automation in the Medical Laboratory Sciences Review Committee of the National Institute of General Medical Sciences; DHEW Publication No. (NIH) 75-762, 1974.

131. Cooks, RG; Beynon, JH; Caprioli, RM; Lester, GR *Metastable Ions*; Elsevier: New York, 1973 (reprinted by ASMS 2004).

132. McFadden, W *Techniques of Combined Gas Chromatography/Mass Spectrometry*: *Applications in Organic Analysis*; Wiley–Interscience: New York, 1973.

133. Costa, E; Holmstedt, B, Eds. *Gas Chromatography-Mass Spectrometry in Neurobiology*; Raven: New York, 1973, ISBN:0911216480.

134. Ahearn, AJ, Ed. *Trace Analysis by Mass Spectrometry*; Academic: New York, 1972.

135. Beckey, H *Field Ionization Mass Spectrometry*; Pergamon: Oxford, U.K., 1971, ISBN:080175570.

136. Leathard, DA; Shurlock, BC *Identification Techniques in Gas Chromatography*; Wiley–Interscience: New York, 1970, ISBN:0471520209.

137. Burlingame, AL; Castagnoli, N, Eds. *Topics in Organic Mass Spectrometry*; Wiley–Interscience: New York, 1970.

138. Ettre, LS; McFadden, WH, Eds. *Ancillary Techniques of Gas Chromatography*; Wiley–Interscience, New York, 1969, ISBN:471246700.

139. Horning EC; Brooks, CJW; Vanden Heuvel, WJA *Gas Phase Analytical Methods for the Study of Steroids*, Vol. 6; Academic: New York, 1968.

Conference Proceedings

The first mass spectrometry conference was organized by the Mass Spectrometry Panel of the Institute of Petroleum and held in Manchester, England, April 20 and 21, 1950. The same group organized another conference held in London, October 29–31, 1953. These two conferences, three years apart, appear to be the predecessor to the triennial International Mass Spectrometry Conferences that began in 1958. The second formal mass spectrometry conference was organized as the NBS Semicentennial Symposium on Mass Spectroscopy in Physics Research held at the National Bureau of Standards in Washington, DC, September 6–8, 1951. The proceedings of this meeting, which appear in a U.S. Government Printing Office publication, is included in this section. The proceedings of all the 16 triennial International Mass Spectrometry Conferences have been published in book form. They are not included in this section; neither are the proceedings of the American Society for Mass Spectrometry's Annual Conference on Mass Spectrometry and Allied Topics nor the proceedings related to meetings dedicated to ion/molecule reactions. Information on the IMSC and ASMS proceedings is included in the **Mass Spectrometry, GC/MS, and LC/MS Journals** section, whereas information on the proceedings of ion/molecule conferences is found in the **Ion/Molecule Reactions** section.

Conference proceedings are important references in that they are often the first time a topic of interest is introduced. These proceedings often have valuable material that never gets into publication in other sources for various reasons and can be used to prevent erroneous turns in undesirable direction while pursuing a goal that involves mass spectrometry. The most infamous of these proceedings is the *Proceedings of the Second Japan–China Joint Symposium on Mass Spectrometry* held in Takarazuka, Japan, September 15–18, 1987 (Matsuda, H; Liang, X-T; Eds.; Bando: Oska, 1987), which contained the presentation (pp 185–188) by Koichi Tanaka (and Ido, Y; Akita, S), Shimadzu Corporation, Kyoto, Japan, on "Soft Laser Desorption" for which Mr. Tanaka became one of the youngest-ever Nobel laureates in chemistry in 2002. This work involved the use of a matrix of metal powder dispersed in glycerol and a pulsed N_2 laser at 337 nm with a coaxial reflectron mass spectrometer. Tanaka reported protonated molecules of carboxypeptidase-A ($M_r = 34,472$) and a spectrum of a pentamer of lysozome ($M_r = 14,306$) at m/z 71,665. This work was later published in *Rapid Communications in Mass Spectrometry* (Tanaka, K; Hiroaki, W; Ido, Y; Akita, A; Yoshida, Y; Yoshida, T) "Protein and Polymer Analyses up to m/z 100,000 by Laser Ionization Time-of-flight Mass Spectrometry" **1988**, *2*(8), 151–153.

Proceedings books are listed separately because oftentimes their titles indicate that they are something other than what they are. These books contain material that pertains to the leading-edge technology at the time of the conference but are often not the best source for gaining knowledge about the subject. Just as edited volumes have the appearance of a camel (a horse built by a committee), the proceedings of a conference are a series of unrelated or, at best, loosely related articles that were presented. Some such proceedings are abstracts collected at the time of the conference. Others are a series of articles, invited by the conference chairs, that may also be peer reviewed.

American Chemical Society Symposium Series books are the results of symposia held at the National ACS Meetings. These books are invited peer-reviewed papers collected into a single volume; however, they are always focused topics. Therefore, reference to these books will appear in this section and also in other sections. Some proceedings references are also duplicated in the **Books of Historical Significance** section.

Interesting history can be gleaned from conference proceedings. As this collection was assembled, a number of volumes edited by Alberto Frigerio (and in some cases, one or more other persons), who worked at the Mario Negri Institute for Pharmacological

Research in Milan, Italy, were discovered. Many of these volumes include the words *Mass Spectrometry*, *Biochemistry*, and *Medicine* in their titles; however, most of the titles are different. There appeared to be no trend in the volume numbers associated with some of the titles, and not all of the titles have associated volume numbers. The last title in the series, *Chromatography and Mass Spectrometry in Nutrition Science and Food Safety*, appeared to have no relation to the other titles. After careful examination, it was determined that these 12 volumes are the proceedings of annual meetings held in Italy between 1972 and 1982 of a conference sponsored by Frigerio's employer. It is also interesting to see the names of the authors of the preface of each of the early volumes, and their selection was done to add credibility to the published works. These preface authors included Frigerio, S. Garattini, E. C. Horning, J. T. Watson, and John Roboz. Beginning with the sixth conference, all the succeeding prefaces were authored by Frigerio.

In addition to the Secondary Ion Mass Spectrometry Meetings (SIMS I–X) whose proceedings are listed in this section, Professor A. Benninghoven (Physikalisches Institut der Universität Münster, Federal Republic of Germany) also organized a series of meetings entitled International Conference on Ion Formation from Organic Solids with the proceedings published as *Ion Formation from Organic Solids (IFOS II–V) Mass Spectrometry of Involatile Material*. These books, like the SIMS Meeting Proceedings, are edited by Benninghoven and are part of the Springer-Verlag Series in Chemical Physics. A listing of the proceedings of the IFOS meetings is not included in this section. Also not included but may be of interest to the mass spectrometrist are the proceedings of the biennial International Symposium on Resonance Ionization Spectroscopy and its Applications, which began in 1982, with the 10th meeting being held in 2000. Details of the 10th meeting were found on the Web at http://www.phys.utk.edu. However, no information was found regarding a 2002 meeting or any subsequent meeting on this topic.

Many small meetings are held each year on mass spectrometry. Some of these never have proceedings reported such as the informal mass spectrometry meeting held annually in Europe. Some will provide abstracts to the participants as is the case with the American Society for Mass Spectrometry's Sanibel Conference, but they often only resemble the material presented because these meetings are designed to be cutting-edge and are specifically intended only for the use of the attendees of the meeting. Information presented at these types of meetings usually finds its way to the peer-reviewed literature or a meeting that has a more formal presentation of the proceedings.

1. Holland, G; Tanner, SD, Eds. *Plasma Source Mass Spectrometry–The New Millennium*, Proceedings of the 7th International Conference on Plasma Source Mass Spectrometry, University of Durham, 10–15 September 2000; Royal Society of Chemistry: Cambridge, U.K., 2003, ISBN:0854048952 (reviewed *JASMS* 14:171).

2. Burlingame, AL; Carr, SA; Baldwin, MA, Eds. *Mass Spectrometry in Biology and Medicine*; Humana: Totowa, NJ, 2000.

3. Shepard, KW, Ed. *Heavy Ion Accelerator Technology*, Eighth International Conference (AIP Conference Proceedings), Vol. 473; American Institute of Physics: New York, 1999.

4. Holland, G; Tanner, SD, Eds. *Plasma Source Mass Spectrometry: New Development and Applications*, International Conference on Plasma Source Mass Spectrometry; Royal Society of Chemistry: Cambridge, U.K., 1999.

5. Gillen G; Lareau, R; Bennett, J; Stevie, FA, Eds. *Secondary Ion Mass Spectrometry: SIMS XI*, Proceedings of the Eleventh International Conference on Secondary Ion Mass Spectrometry, Orlando, Florida, September 7–12, 1997; Wiley: New York, 1998, ISBN:0471978264.

6. Jennings, KR, Ed. *Fundamentals and Applications of Gas-Phase Ion Chemistry*; NATO ASI Series C: Mathematical and Physical Sciences, Vol. 521; Proceedings of the NATO Advanced Study Institute, Grainau, Germany, August 7, 8, 1995; Kluwer Academic: Boston, MA, 1998, ISBN:079235463X.

7. Ens, W; Standing, KG; Chernushevich, IV, Eds. *New Methods for the Study of Biomolecular Complexes*; NATO ASI Series C: Mathematical and Physical Sciences, Vol. 510; Proceedings of the NATO Advanced Research Workshop on New Methods for the Study of Molecular Aggregates, The Lodge at Kananaskis Village, Alberta, Canada, 16–20 June 1996; Kluwer Academic: Boston, MA, 1998.

8. Benninghoven, A; Hagenhoff, B, Eds. *Secondary Ion Mass Spectrometry SIMS X*, Proceedings of the Tenth International Conference on Secondary Ion Mass Spectrometry; Wiley: New York, 1997, ISBN:0471958972.

9. Holland, G; Tanner, SD, Eds. *Plasma Source Mass Spectrometry: New Development and Applications*, Selected Papers from the Fifth International Conference on Plasma Source Mass Spectrometry, University of Durham, 15–20 September 1996; sponsored by Perkin-Elmer Sciex; Royal Society of Chemistry: Cambridge, U.K., 1997, ISBN:0854047271.

10. Caprioli, RM; Malorni, A; Sindona, G, Eds. *Selected Topics in Mass Spectrometry in the Biomolecular Sciences*; NATO ASI Series C: Mathematical and Physical Sciences, Vol. 504; Altavilla-Milicia (PA), Italy, 7–18 July 1996; Kluwer Academic: Boston, MA, 1997, ISBN:0792348494.

11. Caprioli, RM; Malorni, A; Sindona, G, Eds. *Mass Spectrometry in Biomolecular Sciences*; NATO ASI Series C: Mathematical and Physical Sciences, Vol. 475; Lacco Ameno, Ischia, Italy, June 23–July 5, 1993; Kluwer Academic: Boston, MA, 1996, ISBN:0792339460 (reviewed *JASMS* 7:1273).

12. Burlingame, AL; Carr, SA, Eds. *Mass Spectrometry in the Biological Sciences*; Humana: Totowa, NJ, 1996, ISBN:0896033406 (reviewed *JASMS* 7:692).

13. Snyder, AP, Ed. *Biochemical and Biotechnology Applications of Electrospray Ionization Mass Spectrometry*, ACS Symposium Series 619; American Chemical Society: Washington, DC, 1995.

14. Benninghoven, A; Shimizu, R; Werner, HW; Nihei, Y, Eds. *Secondary Ion Mass Spectrometry SIMS IX*, Proceedings of the Ninth International Conference on Secondary Ion Mass Spectrometry, The Hotel Yokohama and The Sangyo-Boeki Center Building, Yokohama, Japan, 7–12 November 1993; Wiley: New York, 1994, ISBN:0471942189.

15. Cotter, RJ, Ed. *Time-of-Flight Mass Spectrometry*; ACS Symposium Series 549; American Chemical Society: Washington, DC, 1994, ISBN:0841227713 (reviewed *JASMS* 7:123).

16. Fenselau, C, Ed. *Mass Spectrometry for the Characterization of Microorganisms*; ACS Symposium Series 549; American Chemical Society: Washington, DC, 1994, ISBN:0841227373 (reviewed *JASMS* 6:1262).

17. Constatin, E, Ed. *Therapeutic Aspects and Analytical Methods in Cancer Research*, Proceedings of the Meeting: Mass Spectrometry in Cancer Research, Proceedings of an Informal Meeting, Strasbourg, France, July 22, 23, 1993; Amudes: Strasbourg Eckbolsheim, France, 1994, ISBN:2906465046.

18. Holland G; Eaton, AN, Eds. *Applications of Plasma Source Mass Spectrometry II*, Proceedings of the 3rd International Conference on Plasma Source Mass Spectrometry, Durham, United Kingdom, 13–18 September 1992; Royal Society of Chemistry: London, 1993, ISBN:0851864651 (reviewed *JASMS* 6:610).

19. Benninghoven, A; Janssen, KTF; Tümpner, J; Werner, HW, Eds. *Secondary Ion Mass Spectrometry: SIMS VIII*, Proceedings of the Eighth International Conference on Secondary Ion Mass Spectrometry; Wiley: Chichester, U.K., 1992, ISBN:0471930644.

20. Standing, KG; Ens, W, Eds. *Methods and Mechanisms for Producing Ions from Large Molecules*; NATO ASI Series B: Physics; Proceedings of the NATO Advanced Research Workshop *Making the Big Ones Fly,* Minaki Lodge, Minaki, Canada, June 24–28, 1990, Vol. 269; Plenum: New York, 1991, ISBN:0306440172.

21. Jennings, KR, Ed. *Fundamentals of Gas-Phase Ion Chemistry*, NATO ASI Series C: Mathematical and Physical Sciences, Proceedings of the NATO Advanced Study Institute, Mont Ste. Odile, France, June 25–July 6, 1990, Vol. 347; Kluwer Academic: Boston, MA, 1991, ISBN:0792314239 (reviewed *JASMS* 4:832).

22. Holland G; Eaton, AN, Eds. *Applications of Plasma Source Mass Spectrometry*, Proceedings on the 2nd International Conference on Plasma Source Mass Spectrometry held at Durham, United Kingdom, 13–18 September 1991; Royal Society of Chemistry: London, 1991 (reviewed *JASMS* 4:428). It should be noted that there were no published proceedings for the first International Conference on Plasma Mass Spectrometry.

23. Brown, MA, Ed. *Liquid Chromatography/Mass Spectrometry: Applications in Agricultural, Pharmaceutical and Environmental Chemistry*; ACS Symposium Series 420; American Chemical Society: Washington, DC, 1990, ISBN:0841217408 (reviewed *JASMS* 4:831).

24. Burlingame, AL; McCloskey, JA, Eds. *Biological Mass Spectrometry*, Proceedings of the 2nd International Symposium of Mass Spectrometry in Health & Life Sciences, San Francisco, California, August 27–31, 1989; Elsevier: Amsterdam, 1990.

25. Benninghoven, A; Werner, HW; Evans, CA; Storms, HA; McKeegan, KD, Eds. *Secondary Ion Mass Spectrometry SIMS VII*, Proceedings of the Seventh International Conference on Secondary Ion Mass Spectrometry; Wiley: New York, 1990, ISBN:0471927384.

26. Jarvis, KE; Gray, AL; Williams, JG, Eds. *Plasma Source Mass Spectrometry*, The Proceedings of the Third Surrey Conference on Plasma Source Mass Spectrometry; Royal Society of Chemistry: London, 1990.

27. McNeal, CJ, Ed. *The Analysis of Peptides and Proteins by Mass Spectrometry*, Proceedings of the 4th Texas Symposium, College Station, Texas, April 17–20, 1988; Wiley: New York, 1988.

28. Slodzian, G; Huber, AM; Benninghoven, A, Eds. *Secondary Ion Mass Spectrometry SIMS VI*, Proceedings of the Sixth International Conference, Palais DES Congre Versailles, France, September 13–18, 1987; Wiley: Chichester, U.K., 1988, ISBN:0471918326.

29. Buchanan, MV, Ed. *Fourier Transform Mass Spectrometry*; American Chemical Society: Washington, DC, 1987.

30. *Proceedings of the 12th Japanese Society for Medical Mass Spectrometry*.

31. Matsuda, H; Liang, X-T; Eds. *Proceedings of the Second Japan–China Joint Symposium on Mass Spectrometry*, Takarazuka, Japan, September 15–18, 1987; Bando: Oska, 1987.

32. Colton, RJ; Benninghoven, A; Simon, DS; Werner, HW, Eds. *Secondary Ion Mass Spectrometry SIMS V*, Proceedings of the Fifth International Conference; Springer Series in Chemical Physics; Springer-Verlag: New York, 1986, ISBN:0387162631.

33. McNeal, CJ, Ed. *Mass Spectrometry in the Analysis of Large Molecules*, Proceedings of the 3rd Texas Symposium, College Station, Texas, April, 1986; Wiley: Chichester, U.K., 1986, ISBN:047191262X.

34. Lyon, PA, Ed. *Desorption Mass Spectrometry: Are SIMS and FAB the Same?*; American Chemical Society: Washington, DC, 1985.

35. Benninghoven, A; Werner, HW; Olano, J; Shimizu, R, Eds. *Secondary Ion Mass Spectrometry SIMS IV*, Proceedings of the Fourth International Conference, Osaka, Japan, November 13–19, 1983; Springer Series in Chemical Physics, Vol. 36; Springer-Verlag: Berlin, 1984.

36. Frigerio, A; Milon, H, Eds. *Chromatography and Mass Spectrometry in Nutrition Science and Food Safety*, Proceedings of the International Symposium on Chromatography and Mass Spectrometry in Nutrition Science and Food Safety, June 19–22, 1983; Analytical Chemistry Symposium Series, Vol. 21; Elsevier: Amsterdam, 1984, ISBN:0444423397.

37. Frigerio, A, Ed. *Chromatography and Mass Spectrometry in Biomedical Sciences, 2*, Proceedings of the International Conference on Chromatography and Mass Spectrometry in Biomedical Sciences, Bordingera, Italy, June 20–30, 1982; Analytical Chemistry Symposium Series, Vol. 14; Elsevier: Amsterdam, 1983, ISBN:0444421548.

38. Schmidt, H-L; Förstel, H; Heinzinger, K *Stable Isotopes*, Proceedings of the 4th International Conference, Jülich, March 23–26, 1981; Analytical Symposia Series, Vol. 11; Elsevier: Amsterdam, 1982, ISBN:0444420762.

39. Benninghoven, A; Giber, J; Laszlo, J; Riedel, M; Werner, HW, Eds. *Secondary Ion Mass Spectrometry Sims III*, Proceedings of the Third International Conference, Technical University, Budapest, Hungary, August 30–September 5, 1981; Springer Series in Chemical Physics, Vol. 19; Springer-Verlag: Berlin, 1982.

40. Frigerio, AF, Ed. *Recent Developments in Mass Spectrometry in Biochemistry, Medicine and Environmental Research, 8*, Proceedings of the 8th International Symposium on Mass Spectrometry in Biochemistry and Medicine, June 18, 19, 1981; Analytical Chemistry Symposium Series, Vol. 8; Elsevier: Amsterdam, 1982.

41. Frigerio, A, Ed. *Recent Developments in Mass Spectrometry in Biochemistry, Medicine and Environmental Research, 7*, Proceedings of the 7th International Symposium on Mass Spectrometry in Biochemistry and Medicine, June 16–18, 1980; Analytical Chemistry Symposium Series, Vol. 7; Elsevier: Amsterdam, 1981.

42. Morris, Ed. *Soft Ionization Biological Mass Spectrometry*, Proceedings of the Chemical Society Symposium on Advances in Mass Spectrometry Soft Ionization Methods, London, July 1980; Heyden: London, 1981, ISBN:0855017066.

43. Frigerio, A; McCarmish, M, Eds. *Recent Developments in Mass Spectrometry in Biochemistry and Medicine, 6*, Proceedings of the 6th International Symposium on Mass Spectrometry in Biochemistry and Medicine, June 21, 22, 1979; Analytical Chemistry Symposium Series, Vol. 4; Elsevier: Amsterdam, 1980.

44. Klein, ER; Klein, PD, Eds. *Stable Isotopes*, Proceedings of the Third International Conference, Oak Brook, Illinois, May 23–26, 1978; Academic: New York, 1979, ISBN: 0124136508.

45. Benninghoven, A; Storms, HA; Evans, CA; Shimizu, R; Powell, RA, Eds. *Secondary Ion Mass Spectrometry SIMS II*, Proceedings of the Second International Conference on Secondary Ion Mass Spectrometry, Stanford University, August 27–31, 1979; Springer Series in Chemical Physics; Springer-Verlag: Berlin, 1979.

46. Frigerio, A, Ed. *Recent Developments in Mass Spectrometry in Biochemistry and Medicine*, Vol. 2, Proceedings of the 5th International Symposium on Mass Spectrometry in Biochemistry and Medicine, Rimini, Italy, June 1978; Plenum: New York, 1979, ISBN:0306402947.

47. Frigerio, A, Ed. *Recent Developments in Mass Spectrometry in Biochemistry and Medicine*, Vol. 1, Proceedings of the 4th International Symposium on Mass Spectrometry in Biochemistry and Medicine, Riva del Garda, Italy, June 1977; Plenum: New York, 1978, ISBN:0306311380.

48. Gross, ML, Ed. *High Performance Mass Spectrometry: Chemical Applications*; ACS Symposium Series 70; American Chemical Society: Washington, DC, 1978.

49. Baillie, TA, Ed. *Stable Isotopes – Applications in Pharmacology, Toxicology and Clinical Research*, Proceedings of an International Symposium on Stable Isotopes, Royal Postgraduate Medical School, London, January 3, 4, 1977; sponsored by the British Pharmacological Society; MacMillan LTD: London, 1978.

50. Jones, CER; Cramers, CA, Eds. *Analytical Pyrolysis*, Proceedings of the Third International Symposium on Analytical Pyrolysis, Amsterdam, September 7–9, 1976; Elsevier: Amsterdam, 1977, ISBN:0444415580.

51. De Leenheer, AP; Roncucci, RR, Eds. *Quantitative Mass Spectrometry in Life Sciences*, Proceedings of the First International Symposium, State University of Ghent, June 16–18, 1976; Elsevier: Amsterdam, 1977, ISBN:0444415572.

52. Frigerio, A; Ghisalberti, EL, Eds. *Mass Spectrometry in Drug Metabolism*, Proceedings of the International Symposium on Mass Spectrometry in Drug Metabolism, Mario Negri Institute for Pharmacological Research, Milan, Italy, June 1976; Plenum: New York, 1977, ISBN:0333217470.

53. Frigerio, A, Ed. *Advances in Mass Spectrometry in Biochemistry and Medicine*, Vol. II, Proceedings of the 3rd International Symposium on Mass Spectrometry in Biochemistry and Medicine, Mario Negri Institute for Pharmacological Research, Milan, Italy, June 1975; Spectrum Publications: New York, 1977, ISBN:0893350087.

54. Frigerio, A, Ed. *Advances in Mass Spectrometry in Biochemistry and Medicine*, Vol. I, Proceedings of the 2nd International Symposium on Mass Spectrometry in Biochemistry and Medicine, Mario Negri Institute for Pharmacological Research, Milan, Italy, June 1974; Halsted: New York, 1976, ISBN:0470281219.

55. *Frigerio, A; Castagnoli, N, Jr., Eds. *Mass Spectrometry in Biochemistry and Medicine*, Monographs of the Mario Negri Institute for Pharmacological Research, Milan, Italy; Raven: New York, 1974, ISBN:0911216537.

56. Mamer, OA; Mitchell, WJ; Scriver, CR, Eds. *Application of Gas Chromatography – Mass Spectrometry to the Investigation of Human Disease*, The Proceedings of a Workshop, Montreal, Montreal Children's Hospital and Royal Victoria Hospital, May 30, 31, 1973; McGill University–Montreal Children's Hospital Research Institute: Montreal, 1974.

57. Frigerio, A, Ed. *Proceedings of the International Symposium on Gas Chromatography Mass Spectrometry*, Isle of Elba, Italy, 17–19 May 1972; Tamburini: Milano, Italy, 1972.

58. Reed, RI, Ed. *Recent Topics in Mass Spectrometry*, NATO Study Institute of Mass Spectrometry, Lisbon, Spain, August 1969; Gordon and Breach: New York, 1971, ISBN:0677148003.

59. Ogata, K; Hayakawa, T, Eds. *Recent Developments in Mass Spectroscopy*, Proceedings of the International Conference on Mass Spectroscopy, Kyoto, Japan, September 8–12, 1969; University of Tokyo Press: Tokyo, 1970.

60. Price, D; Williams, JE, Eds. *Time of Flight Mass Spectrometry* (1st proceeding), including a 14-page bibliography; Pergamon: London, 1969, ISBN:0080134440; all subsequent proceedings entitled *Dynamic Mass Spectrometry*, Vol. 1, 1970 (2nd) ISBN:0855010339; Vol. 2, 1971 (3rd) ISBN:0855010541; Vol. 3, 1972 (4th) Price, D, Ed., ISBN:0855010614, published by Heyden: London; Vol. 4, 1976 (5th), Price, D; Todd, JFJ, Eds., ISBN:0471259640; Vol. 5, 1978 (6th) ISBN:0471259667; Vol. 6, 1981 (7th) ISBN:0471259XXX, published by Wiley after the acquisition of Heyden: London, as the proceedings of the first seven European Time-of-Flight Symposia. It appears that this symposium then became triennial with the 8th being held in 1983, the 9th in 1986, and the 10th in 1989. All symposia were held in Salford, United Kingdom.

61. Brymner, R; Penney, JR, Eds. *Mass Spectrometry*, Proceedings of the Symposium on Mass Spectrometry, Enfield College of Technology, July 5, 6, 1967; Chemical: New York, 1969.

62. Reed, RI, Ed. *Modern Aspects of Mass Spectrometry*, Proceedings of the 2nd NATO Advanced Study Institute of Mass Spectrometry on Theory, Design, and Applications, July 1966, University of Glasgow, Glasgow, Scotland; Plenum: New York, 1968.

63. Reed, RI, Ed. *Mass Spectrometry*, Proceedings of the 1st NATO Advanced Study Institute on Theory, Design, and Applications, Glasgow, Scotland, August 1964; Academic: London, 1965.

64. Hinteberger, H, Ed. *Nuclear Masses and their Determination*, Proceedings of the Conference "Max-Planck-Institute für Chemie" Mainz, 10–12 July 1955; Pergamon: London, 1957.

65. Smith, ML, Ed. *Electromagnetically Enriched Isotopes and Mass Spectrometry*, Proceedings of the Harwell Conference, Sept. 13–16, 1955; Butterworth: London, 1956.

66. Blears, J (Chairman, Mass Spectrometry Panel, Institute of Petroleum) *Applied Mass Spectrometry*, a report of a conference organized by The Mass Spectrometry Panel of The Institute of Petroleum, London, 29–31 October 1953; The Institute of Petroleum: London, 1954.

67. Hipple, JA; Aldrich, LT; Nier, AOC; Dibeler, VH; Mohler, FL; O'Dette, RE; Odishaw, H; Sommer, H (Mass Spectroscopy Committee) *Mass Spectrometry in Physics Research*, National Bureau of Standards Circular 522; United States Government Printing Office: Washington, DC, 1953.

68. *Mass Spectrometry,* a report of a conference organized by The Mass Spectrometry Panel of The Institute of Petroleum, London, April 20, 21, 1950; The Institute of Petroleum: London, 1952.

Interpretation Books

The seminal book for the interpretation of EI mass spectra is the McLafferty book (Interpretation 7). This book is extremely valuable but may be too advanced for a beginner trying to self-teach. The beginner should try to start with either the McLafferty 2nd edition book (Interpretation 16) or the book by Shrader (Interpretation 19), both of which have been out of print for some time but can be found on the used-book Web sites.

The books by Budzikiewicz et al. (Interpretation 23, 27, and 28) were the first books written using mechanisms in organic reactions to describe the fragmentation of energetic ions produced by electron ionization. The information in these books still has a great deal of relevance to the subject. All of the interpretation books listed in this section pertain primarily to odd-electron molecular ions formed by the EI process. The ions formed in LC/MS are predominately protonated molecules, which are even-electron ions. Even-electron ions will most often fragment to produce other even-electron ions, which requires the breaking of more than one bond in the ion. Many of the neutral losses in even-electron ion fragmentation and odd-electron ion fragmentation are the same.

One important note on Interpretation 4: Just as book titles can be similar (Technique 38 and 52), people can have the same or similar names that can result in confusion. The author of Interpretation 4, Terrence A. Lee, a Department of Chemistry faculty member at Middle Tennessee State University in Murfreesboro, Tennessee, should not be confused with Terry Lee, a noted researcher in mass spectrometry of biological substances at the Beckman Research Institute, City of Hope/Division of Immunology in Duarte, California.

1. Smith, RM *Understanding Mass Spectra: A Basic Approach*, 2nd ed.; Wiley: Hoboken, NJ, 2004, ISBN:047142949X (reviewed *JASMS* 16:792). Smith, RM *Understanding Mass Spectra: A Basic Approach*; Busch, KL, Tech. Ed.; Wiley: New York, 1999, ISBN:0471297046 (reviewed *JASMS* 11:664).

2. Snyder, AP *Interpreting Protein Mass Spectra: A Comprehensive Resource*; Oxford: New York, 2000, ISBN:0842135716 (reviewed *JASMS* 13:107).

3. Barker, J *Mass Spectrometry: Analytical Chemistry by Open Learning*, 2nd ed.; Ando, DJ, Ed.; Wiley: Chichester, U.K., 1999, ISBN:0471967645 (Davis, R; Frearson, MJ, 1st ed., 1987, ISBN:0471913898; reviewed *JASMS* 4:831).

4. Lee, TA *A Beginner's Guide to Mass Spectral Interpretation*; Wiley: Chichester, U.K., 1998, ISBN:047197628 hc; ISBN:0471976296 pbk (reviewed *JASMS* 9:852).

5. *Watson, JT *Introduction to Mass Spectrometry*, 3rd ed.; Lippincott-Raven: Philadelphia-New York, 1997.

6. Splitter, JS; Tureček, F, Eds. *Applications of Mass Spectrometry to Organic Stereochemistry*; VCH: New York, 1994, ISBN:089573303X (reviewed *JASMS* 6:152).

7. McLafferty, FW; Tureček, F *Interpretation of Mass Spectra*, 4th ed.; University Science: Mill Valley, CA, 1993, ISBN:0935702253 (reviewed *JASMS* 5:949).

8. ЭАИКИН, ВГ; МИКАЯ, АИ *ХИМИЧЕСКИЕ МЕТОДЫ В МАСС- СПЕКТРОМЕТРИИ ОРГАНИЧЕСКИХ СОЕДИНЕНИЙ*; Издательство "Hayka": Moscow, 1987.

9. ТЕРЕНТЬЕВ, ПБ; СТАНКЯВИЧЮС, АП Масс-спектрометрический анализ биологически активных азотистых оснований; Иэдательство: Мокслас, 1987.

10. ВУЛьФСОН, НС; ЭАИКИН, ВГ; МИКАЯ, АИ *МАСС-СПЕКТРОМЕТРИЯ ОРГАНИЧЕСКИХ СОЕДИНЕНИЙ*; Издательство "Наука": Moscow, 1986.

11. Porter, QN *Mass Spectrometry of Heterocyclic Compounds*, 2nd ed.; Wiley–Interscience: New York, 1985.

12. McLafferty, FW; Venkataraghavan, R *Mass Spectral Correlations*, 2nd ed.; American Chemical Society: Washington, DC, 1982.

13. Sklarz, B, Ed. *Mass Spectrometry of Natural Products*, plenary lecturers presented at the International Mass Spectrometry Symposium on Natural Products, Rehovot, Israel, 28 August–2 September 1977; Pergamon: Oxford, U.K., 1978.

14. Levsen, K *Fundamental Aspects of Organic Mass Spectrometry*; Verlag Chemie: Weinheim, Germany, 1978.

15. DeJongh, DC *Interpretation of Mass Spectra*; ACS Audio Series; American Chemical Society: Washington, DC, 1975; 6 audio cassette tapes, 158 pp.

16. McLafferty, FW *Interpretation of Mass Spectra*, 2nd ed.; Benjamin: Reading, MA, 1973.

17. Hamming, MG; Foster, NG *Interpretation of Mass Spectra of Organic Compounds*; Academic: New York, 1972.

18. Hill, HC *Introduction to Mass Spectrometry*, 2nd ed.; Heyden: London, 1972. (1st ed., 1966; 2nd printing, 1969; 2nd ed. revised by AG Loudon; 1st ed. translated into German, Italian, and Japanese; Hilson C. Hill died in 1967).

19. Shrader, SR *Introductory Mass Spectrometry*, Allyn and Bacon: Boston, MA, 1971.

20. Seibl, J *Massenspektrometrie*; Akademische Verlagsgesellschaft: Frankfurt, Germany, 1970.

21. Beynon, JH; Saunders, RA; Williams, AE *The Mass Spectra of Organic Molecules*; Elsevier: Amsterdam, 1968.

22. Polyakova, AA; Khmel'nitskii, RA *Introduction to Mass Spectrometry of Organic Compounds*; Schmorak, J, Translator; Israel Program For ScientificTranslations: Jerusalem, Israel, 1968 (original Russian language edition, *Vvedenie V Mass Spektrometriyu Organicheskikh Soedinenii*; Izdatel'stvo "Khimya": Moskva-Leningrad, 1966).

23. Budzikiewicz, H; Djerassi, C; Williams, DH *Mass Spectrometry of Organic Compounds*; Holden-Day: San Francisco, CA, 1967.

24. Reed, RI *Applications of Mass Spectrometry to Organic Chemistry*; Academic: New York, 1966.

25. Spiteller, G *Massenspektrometrische Strukturanalyse Organischer Verbindungen*; Verlag Chemie: Weinheim, Germany, 1966.

26. Quayle, A; Reed, RI Interpretation of Mass Spectra. In *Interpretation of Organic Spectra*; Mathieson, DW, Ed.; Academic: New York, 1965.

27. Budzikiewicz, H; Djerassi, C; Williams, DH *Structure Elucidation of Natural Products by Mass Spectrometry*, Vol. I *Alkaloids*; Vol. II *Steroids, Terpenoids, Sugars, and Miscellaneous Natural Products*; Holden-Day: San Francisco, CA, 1964.

28. Budzikiewicz, H; Djerassi, C; Williams, DH *Interpretation of Mass Spectra of Organic Compounds*; Holden-Day: San Francisco, CA, 1964.

29. McLafferty, FW, Ed. *Mass Spectra of Organic Ions*; Academic: New York, 1963.

30. McLafferty, FW Mass Spectrometry. In *Determination of Organic Structures by Physical Methods*, Vol. II; Nachod, FC; Phillips, WD, Eds.; Academic: New York, 1962.

Ion/Molecule Reactions

The ion/molecule reaction is as important to mass spectrometry as any other aspect of ionization. These phenomena have been extensively studied. The books listed in this section are those relating to ion/molecule reactions that are the results of individual studies and symposia relating to the field.

1. Ng, C-Y; Babcock, LM, Eds. *Advances in Gas Phase Ion Chemistry*, Vol. 4; Jai Press: Greenwich, CT, 1999, ISBN:0762304383 (Vol. 1, 1992, ISBN:1559383313; Vol. 2, 1996, ISBN:1559387033; Vol. 3, 1998, ISBN:0762302046; Vols. 1, 2 reviewed *JASMS* 7:1178).

2. Whelan, CT; Walters, HRJ, Eds. *Coincidence Studies of Electron and Photon Impact Ionization*; Physics of Atoms and Molecules Series; Burke, PG; Kleinpoppen, H, Series Eds.; Plenum: New York, 1997, ISBN:0306456893.

3. Ng, C-Y; Baer, M; Powis, I, Eds. *Unimolecular and Bimolecular Ion-Molecule Reaction Dynamics*; Wiley: New York, 1994, ISBN:0471938319.

4. Ng, C-Y; Baer, M, Eds. *State-Selected and State-to-State Ion-Molecule Reaction Dynamics*, Part 1: *Experiment*; Part 2: *Theory*; Vol. 82 in *Advances in Chemical Physics*; Wiley: New York, 1992, ISBN:0471532584.

5. Farrar, JM; Saunders, WH, Eds. *Techniques for the Study of Ion-Molecule Reactions*, Vol. 20 *Techniques of Chemistry*; Wiley: New York, 1988, ISBN:0471848123.

6. Bowers, MT, Ed. *Gas Phase Ion Chemistry*, Vols. 1, 2, 1979; Vol. 3, 1984; Academic: New York.

7. Franklin, JL, Ed. *Benchmark Papers in Physical Chemistry and Chemical Physics*, Vol. 3 *Ion-Molecule Reactions, Part I: The Nature of Collisions and Reactions of Ions with Molecules and Ion-Molecule Reactions*; *Part II: Elevated Pressures and Long Reaction Times*; Dowden, Hutchingson & Ross: Stroudsburg, PA, 1979.

8. Ausloos, P, Ed. *Kinetics of Ion-Molecule Reactions*, Proceedings of the NATO Advanced Study Institute on Kinetics of Ion-Molecule Reactions, Biarritz, France, September 4–15, 1978, Vol. 40; published in conjunction with NATO Scientific Affairs Division; Plenum: New York, 1979.

9. Ausloos, P, Ed. *Interactions Between Ions and Molecules*, Proceedings of the NATO Advanced Study Institute on Kinetics of Ion-Molecule Reactions, La Baule, France, 1974; published in conjunction with NATO Scientific Affairs Division; Plenum: New York, 1975.

10. Lias, SG; Ausloos, PJ *Ion-Molecule Reactions: Their Role in Radiation Chemistry*; American Chemical Society: Washington, DC, 1975.

11. Franklin, JL, Ed. *Ion-Molecule Reactions*, Vols. 1, 2; Plenum: New York, 1972.

12. McDaniel, EW; Chermák, V; Dalgarno, A; Ferguson, EE; Friedman, L *Ion-Molecule Reactions*; Wiley–Interscience: New York, 1970.

13. Knewstubb, PF *Mass Spectrometry and Ion-molecule Reactions*; Cambridge University Press: Cambridge, U.K., 1969.

14. Moiseiwitsch, BL; Smith, SJ *Electron Impact Excitation of Atoms*; National Standard Reference Data Series; National Bureau of Standards; Astin, AV, Director; Smith, CR, Director; United States Department of Commerce: Washington, DC, August, 1968; (reprinted from *Rev. Mod. Phys.* **1968** [April], *40*[2]).

15. Ausloos, PJ, Ed. *Ion-Molecule Reactions in the Gas Phase*; ACS Advances in Chemistry Series 58; American Chemical Society: Washington, DC, 1966.

16. McDaniel, EW *Collision Phenomena in Ionized Gases*; Wiley: New York, 1964.

Books of Historical Significance[*]

Based on an 1886 paper (*Berl. Ber.* **1886**, *39*, 691) by Eugene Goldstein (German physicist, 1850–1930) reporting the discovery of luminous rays emerging as straight lines from holes in a metal disc used as a cathode in a discharge tube (he called the rays *Kanalstrahlen*: canal rays) and the confirmation by Wilhelm Carl Werner Otto Fritz Franz Wien (German Nobel laureate in physics, 1911, 1864–1928) that Jean Baptiste Perrin's (French physicist, 1870–1942) 1895 postulation that the rays were associated with positive charge by studying their deflection in electric and magnetic fields (*Verh. Phys. Ges.* **1898**, *17*, 1898; *Ann. Physik.* **1898**, *65*, 440; *Ann. Phys. Leipzig* **1902**, *8*, 224), the field of mass spectrometry developed into a science between 1911 and 1925. This development was due to the results of the experiments conducted by the three founding fathers of mass spectrometry: Joseph John Thomson (English Nobel laureate in physics, 1906, 1856–1940); Francis William Aston (English Nobel laureate in chemistry, 1922, 1877–1945; Aston was an associate of Thomson in the Cavendish Laboratory in Manchester, England); and Arthur Jeffery Dempster (1886–1950, Canadian-American physics professor, University of Chicago).

In his 1968 book, Roboz (Reference 86) lists 20 selected papers for those wanting to learn the history of mass spectrometry through original references (Chapter 14, p 490). Of these papers, five were authored by Aston, four by Dempster, and two by Thomson. Another three were authored by William R. Smythe (U.S. scientist) and two by Kenneth Bainbridge (U.S. physicist, 1904–1906, Director of the Trinity test—the first test explosion of the atomic bomb), who also were early pioneers in mass spectrometry.

The first half of the 20th century resulted in three major mass spectrometry book titles, each having two editions. These were the Thomson (one title, two editions) and the Aston (two titles, two editions each) books, which were published between 1913 and 1942 (the year of the first announcement of a commercial mass spectrometer by Westinghouse). No books other than the three titles by Louis Cartan (France) and the titles by Simion Bauer (United States) and Robert M. Dowben (Norway) on this subject appeared until Ewald and Hintenberger published a German-language book in 1952 that was translated into English in 1962. This was followed in 1953 by Barnard's *Modern Mass Spectrometry* and Kenneth T. Bainbridge's 200-plus-page book published as Part IV of E. Sergè's *Experimental Nuclear Physics*, Vol. 1. In 1954, *Methuen's Monograph on Chemical Subjects*, authored by Robertson; and Inghram and Hayden's *A Handbook of Mass Spectroscopy* were published. Duckworth's first edition of *Mass Spectroscopy* appeared in 1958. There were other books published in the 1950s of interest to the mass spectrometrist, but they were either proceedings of conferences or more related to physics as was the case with the Massey and Loeb books of the 1930s and 1940s. The Library of Congress Catalogue lists a number of "book" titles from the 1930s, which are believed to be nothing more than separately bound reprints of individual articles. The 1960s saw a significant number of seminal books published by notables such as Biemann; Beynon; and Budzikiewicz, Djerassi, and Williams, just to name a few. Over the last 30 years, with the large number of published conference and American Chemical Society symposia proceedings and special topics books, the number of mass spectrometry titles has grown to well over 500.

---

[*] Historical references are in the Bibliography section of the Kiser book (Reference 90) and in the Information and Data chapter of the Roboz book (Reference 86).

1. Davis, EA; Falconer, IJ *J. J. Thomson and the Discovery of the Electron*; Taylor & Francis: London, 1997.

2. Dahl, PF *Flash of the Cathode Rays: A History of JJ Thomson's Electron*; American Institute of Physics: Philadelphia, PA, 1997.

3. Nachod, FC; Zuckerman, JJ; Randall, EW, Eds. *Determination of Organic Structures by Physical Methods*, Vol. 6; Academic: New York, 1976.

4. Frigerio, A, Ed. *Mass Spectrometry in Biochemistry and Medicine*, Vols. 1, 2 *Advances in Mass Spectrometry in Biochemistry and Medicine*, 1974, 1975; *Mass Spectrometry in Drug Metabolism*, 1976; Vols. 1, 2, 6, 7 and Vol. 8 *Recent Developments in Mass Spectrometry in Biochemistry and Medicine*, 1977–1980 and 1982; a series of books published by Plenum: New York and Elsevier: Amsterdam as the proceedings of a meeting organized by the Mario Negri Institute for Pharmacological Research in Milan, Italy.

5. Biennial Specialist Periodical Reports: *Mass Spectrometry, A Review of the Recent Literature Published between July 19XX and June 19XX+2*, Vol. 1 (1968–1970), 1971; Vol. 2 (1970–1972), 1973, Williams, DH, Ed.; Vol. 3 (1972–1974), 1975; Vol. 4 (1974–1976), 1977; Vol. 5 (1976–1978), 1979; Vol. 6 (1978–1980), 1981; Vol. 7 (1980–1982), 1984, Johnstone, RAW, Ed.; Vol. 8 (1982–1984), 1985; Vol. 9 (1984–1986), 1987; Vol.10 (1986–1988), 1989, Rose, ME, Ed.; Royal Society of Chemistry (formerly Chemical Society): Cambridge, U.K.

6. Ogata, K; Hayakawa, T, Eds. *Recent Developments in Mass Spectroscopy*, Proceedings of the International Conference on Mass Spectroscopy, Kyoto, Japan, September 8–12, 1969; University of Tokyo Press: Tokyo, 1970.

7. Price, D; Williams, JE, Eds. *Time of Flight Mass Spectrometry* (1st proceeding), 1969; all subsequent proceedings entitled *Dynamic Mass Spectrometry*, Vol. 1, 1970 (2nd); Vol. 2, 1971 (3rd); Vol. 3, 1972 (4th) Price, D, Ed.; Vol. 4, 1976 (5th) Price, D; Todd, JFJ, Eds.; Vol. 5, 1978 (6th); Vol. 6, 1981 (7th); published by Heyden: London as the proceedings of the seven European Time-of-Flight Symposia.

8. Massey, HSW; Burhop, EHS; Gilbody, HB *Electronic and Ionic Impact Phenomena*, 2nd ed., Vol. I *Electron Collisions with Atoms*, 1969; Vol. II *Electron Collisions with Molecules – Photoionization*, 1969; Vol. III *Slow Collisions of Heavy Particles*, 1971; Vol. IV *Recombination and Fast Collisions of Heavy Particles*, 1974; Vol. V *Slow Positron and Muon Collisions – Notes on Recent Advances*, 1974; Oxford University Press: London.

9. Kientiz, H *Massenspektrometrie*; Verlag Chemie: Weinheim, Germany, 1968.

10. Reed, RI, Ed. *Modern Aspects of Mass Spectrometry*, Proceedings of the 2nd NATO Advanced Study Institute of Mass Spectrometry on Theory, Design, and Applications, July 1966, University of Glasgow, Glasgow, Scotland; Plenum: New York, 1968.

11. Blauth, EW *Dynamic Mass Spectrometers* (translated from German); Elsevier: Amsterdam, 1966.

12. Jayaram, R *Mass Spectrometry: Theory and Applications*; Plenum: New York, 1966.

13. Mead, WL, Ed. *Advances in Mass Spectrometry*, Vol. 3; Pergamon: New York, 1966.

14. Thomson, GP *J. J. Thomson: Discoverer of the Electron*; Doubleday Anchor: Garden City, NY, 1966.

15. Thomson, GP *J. J. Thomson and the Cavendish Laboratory in His Day*; Doubleday: New York, 1965.

16. Reed, RI, Ed. *Mass Spectrometry*, Proceedings of the 1st NATO Advanced Study Institute of Mass Spectrometry on Theory, Design, and Applications; Academic: London, 1965.

17. Loeb, LB *Electric Coronas: Their Basic Physical Mechanisms*; University of California Press: Berkeley, CA, 1965.

18. Mott, NF; Massey, HSW *The Theory of Atomic Collisions*, 3rd ed.; Oxford University Press: London, 1965; 1st ed., 1933; 2nd ed., 1949.

19. Brunnee, C; Voshage, H *Massenspektrometrie*; K. Thiemig: Munich, 1964.

20. Budzikiewicz, H; Djerassi, C; Williams, DH *Interpretation of Mass Spectra of Organic Compounds*; Holden-Day: San Francisco, CA, 1964.

21. McDowell, CA, Ed. *Mass Spectrometry*; McGraw-Hill: New York, 1963 (reprinted by Robert E. Krieger: Huntington, NY, 1979).

22. Elliott, RM, Ed. *Advances in Mass Spectrometry*, Vol. 2; Pergamon: New York, 1963.

23. Biemann, K *Mass Spectrometry: Organic Chemical Applications*; McGraw-Hill: New York, 1962 (reprinted by ASMS, 1998).

24. *Bibliography on Mass Spectrometry 1938–1957 Inclusive* prepared by the Intelligence and Interactive Section, Research Department, Associated Electrical Industries (Manchester) Limited (formerly Metropolitan-Vickers Electrical Company Limited); Pergamon: London, 1961. First published in *Advances in Mass Spectrometry;* Pergamon: London, 1959. First compiled in 1948, the first supplement covered the period of July 1948 through August 1950 and was published along with the orginal in *Mass Spectrometry*, A Report of a Conference Organized by the Mass Spectrometry Panel of the Institute of Petroleum and held April 20, 21, 1950, in Manchester as a bibliography covering the period January 1938 through November 1950; Institute of Petroleum: 1952; the second supplement covered the period of September 1950 through June 1953 and appeared as an appendix to *Applied Mass Spectrometry* (1953 Conference); Institute of Petroleum: London, 1954. Two more supplements covering the intervening periods to 1957 were combined with the original compilation and the first two supplements to comprise the bibliography that appears in the reference and in Vol. 1 of *Advances in Mass Spectrometry* (1958 Conference). The fifth supplement covering the period of January 1958 through December 1960 was first published in *Advances in Mass Spectrometry*, Vol. 2 (1961 Conference); Pergamon: 1963. The sixth supplement covering January 1961 through December 1962 was first published in *Advances in Mass Spectrometry*, Vol. 3 (1964 Conference); Pergamon: 1966. The fifth and sixth supplements were published by Pergamon as individual Supplements 1 and 2 to this reference. In the Editor's Forward in *Advances in Mass Spectrometry*, Vol. 4, Institute of Petroleum: London, 1968, it is stated that because of the United Kingdom formation, the Mass Spectrometry Data Center at Aldermaston, Berks, England, which as of then published the *Mass Spectrometry Bulletin*, that subsequent supplements to the AEI bibliography would no longer be prepared and thus no longer appear as part of the Proceedings of the International Mass Spectrometry Meetings. The Data Center was charged with keeping abreast of the literature from 1964.

25. Beynon, JH *Mass Spectrometry and Its Applications to Organic Chemistry*; Elsevier: Amsterdam, 1960 (reprinted by ASMS, 1999).

26. Waldron, JD, Ed. *Advances in Mass Spectrometry*, Vol. 1; Pergamon: New York, 1959.

27. Duckworth, HE *Mass Spectroscopy*; Cambridge: London, 1958.

28. Loeb, LB *Static Electrification*; Springer-Verlag: Berlin, 1958.

29. Rieck, GR *Einführung in die Massenspektroskopie* (translated from Russian); VEB Deutscher Verlag der Wissebschaften: Berlin, 1956.

30. Smith, ML, Ed. *Electromagnetically Enriched Isotopes and Mass Spectrometry*, Proceedings of the Harwell Conference, September 13–16, 1955: Butterworth; London, 1956.

31. Loeb, LB *Basic Processes of Gaseous Electronics*; University of California Press: Berkeley, CA, 1955 (revised and reprinted as the 2nd edition in 1960 with Appendix I).

32. Robertson, AJB *Mass Spectrometry: Methuen's Monographs on Chemical Subjects*; Wiley: New York, 1954.

33. Blears, J (Chairman, Mass Spectrometry Panel, Institute of Petroleum) *Applied Mass Spectrometry*, A Report of a Conference Organized by The Mass Spectrometry Panel of The Institute of Petroleum, London, 29–31 October 1953; The Institute of Petroleum: London, 1954.

34. Inghram, MG; Hayden, RJ *A Handbook on Mass Spectroscopy*, Nuclear Science Report No. 14; National Academy of Science, National Research Council Publication 311: Washington, DC, 1954 (manuscript completed in April 1952 and intended as a chapter in a proposed handbook on nuclear instruments and techniques for the Subcommittee on Nuclear Instruments and Techniques of the Committee on Nuclear Science of the National Research Council that was never completed).

35. Barnard, GP *Modern Mass Spectrometry*; American Institute of Physics: London, 1953.

36. Bainbridge, KT "Part V: Charged Particle Dynamics and Optics, Relative Isotopic Abundances of the Elements, Atomic Masses" in *Experimental Nuclear Physics*, Vol. 1; Sergè, E, Ed.; Wiley: New York, 1953; pp 559–767.

37. Hipple, JA; Aldrich, LT; Nier, AOC; Dibeler, VH; Mohler, FL; O'Dette, RE; Odishaw, H; Sommer, H (Mass Spectroscopy Committee) *Mass Spectrometry in Physics Research*, National Bureau of Standards Circular 522; United States Government Printing Office: Washington, DC, 1953.

38. Dowben, RM *Mass Spectrometry*; Joint Establishment for Nuclear Energy Research: Kjeller per Lillestrøm: Norway, 1952; 74 pp; diagrams.

39. *Mass Spectrometry*, A Report of a Conference Organized by The Mass Spectrometry Panel of The Institute of Petroleum, London, 20, 21 April 1950; The Institute of Petroleum: London, 1952.

40. Ewald, H; Hintenberger, H *Methoden und Anwendunyngen der Massenspektroskopie*; Verlag Chemie: Weinheim, Germany, 1952 (English translation by USAEC, Translation Series AEC-tr-5080; Office of Technical Service: Washington, DC, 1962).

41. Massey, HSW; Burhop EHS *Electronic and Ionic Impact Phenomena*; Oxford University Press: London, 1952; 2nd printing, 1956.

42. Massey, HSW *Negative Ions*, 2nd ed.; Cambridge at the University: London, 1950; 1st ed., 1938.

43. Aston, FW *Mass Spectrometry and Isotopes*, 2nd ed.; Edward Arnold: London, 1942; 1st ed., 1933.

44. Loeb, LB *Fundamental Processes of Electrical Discharge in Gases*; Wiley: New York, 1939.

45. Cartan, L *L'optique des Rayons Positifs et ses Applications à la Spectrographie de Masse*; Hermann & Cie: Paris, 1938; p 80.

46. Thibaud, J; Cartan, L; Comparat, P *Quelques Techniques Actuelles en Physique Nucléaire*, subtitle that appears to be the contents *Méthode de la Trochoïde: Électrons Positifs: Spectrographie de Masse: Isotopes: Compteurs de Particules à Amplification Linéaire Compteurs de Geiger et Müller*; Hermann & Cie: Paris, 1938; p 276.

47. Cartan, L *Spectrographie de Masse: Les Iotopes et Leurs Masses*; No. 550 in the series *Actualités Scientifiques et Industrielles*; No. VI in the subseries *Exposés de Physique Atomique Expérimentale*; Preface by Maurice de Broglie; Hermann & Cie: Paris, 1937; pp 3–90, pbk.

48. Bauer, SH *A Mass Spectrograph: Products and Processes of Ionization in Methyl Chloride*; 1935; iv, 51 pp; illustrations.

49. Aston, FW *Isotopes*, 2nd ed.; Edward Arnold: London, 1924; 1st ed., 1922.

50. Thomson, JJ *Rays of Positive Electricity and their Application to Chemical Analysis*, 2nd ed.; Longmans Green: London, 1921; 1st ed., 1913.

Collections of Mass Spectra in Hardcopy

There have been many collections of mass spectra that have come and gone. In a 1985 monograph (A Guide To, And Commentary On, The Published Collection and Literature of Mass Spectral Data) published by VG Analytical (the United Kingdom mass spectrometry company now known as Micromass/Waters), 33 separate collections were referenced. In 1974 and 1978, the American Society for Mass Spectrometry published the 1st and 2nd editions of *A Guide to Collections of Mass Spectral Data*. These editions include 24 and 30 references, respectively. None of these collections has been lost. They have all been consolidated into either the National Institute of Standards and Technology (NIST) Mass Spectral Database or the Wiley Registry of Mass Spectral Data, or both. Some of these collections are not currently available in an electronic format, or the electronic format is only of abbreviated spectra that ranges from a minimum of 16 to a maximum of 50 mass spectral peaks (Collections 7). Hardcopy volumes are somewhat less valuable than electronic versions. The Cornu collection is even less valuable because it is a tabular listing of the 10 most intense peaks.

1. Adams, RP *Identification of Essential Oil Components by Gas Chromatography/Quadrupole Mass Spectroscopy*, 3rd ed.; Allured: Carol Stream, IL, 2001, ISBN:0931710855 pbk; 456 pp. This is the third edition on mass spectra and retention times of common components in plant essential oils. Unlike previous two editions (QIT), these 1,606 spectra were acquired using a transmission quadrupole mass spectrometer. The book includes the ion trap spectra of the same 1606 compounds.

2. Makin, HLJ; Trafford, DJH; Nolan, J *Mass Spectra and GC Data of Steroids: Androgens and Estrogens*; Wiley: Chichester, U.K., 1999.

3. Newman, R; Gilbert, MW; Lothridge, K *GC-MS Guide to Ignitable Liquids*; CRC: Boca Raton, FL, 1998.

4. ‡Vickerman, JC; Briggs, D; Henderson, A, Eds. *The Wiley Static SIMS Library*; Wiley: New York, 1996.

5. ‡Adams, RP *Identification of Essential Oil Compounds by Gas Chromatography/Mass Spectrometry*; Allured: Carol Stream, IL, 1995, ISBN:0931710421; 1211 spectra (reviewed *JASMS* 8:671).

6. ‡Mills, T, III; Roberson, JC *Instrumental Data For Drug Analysis*, 2nd ed., Vols. 1–5, Vol. 6, Vol. 7; CRC: Boca Raton, FL, 1993 (originally published by Elsevier: New York, 1987–1992).

7. ‡Pfleger, K; Maurer, WW; Weber, A *Mass Spectral and GC Data of Drugs, Pollutants, Pesticides and Metabolites*, 2nd ed., 3-volume set; VCH: New York, 1992; 4th volume was added in 1999; 4-volume set, ISBN:3527297936.

8. Hites, RA *CRC Handbook of Mass Spectra of Environmental Contaminants*, 2nd ed.; CRC: Boca Raton, FL, 1992, ISBN:0873715349; 533 spectra (reviewed *JASMS* 5:598); 1st ed., 1985, ISBN:084930537; 394 spectra.

9. McLafferty, FW; Stauffer, DB *Important Peak Index of the Registry of Mass Spectral Data*, 3 volumes; Wiley: New York, 1991, ISBN:0471552704 (reviewed *JASMS* 4:82).

10. *The Eight Peak Index of Mass Spectra*, 4th ed.; Royal Society of Chemistry: Cambridge, U.K., 1991.

11. ‡McLafferty, FW; Stauffer, DB *The Wiley/NBS Registry of Mass Spectral Data*, 7 volumes; Wiley: New York, 1989, ISBN:0471628867; 133,000 spectra.

12. Pace-Asciak, CR *Mass Spectra of Prostaglandins and Related Products*; Lippincott-Raven: Philadelphia, PA, 1989, ISBN:0881674745.

13. Stemmler, EA; Hites, RA *Electron Capture Negative Ion Mass Spectra of Environmental Contaminants and Related Compounds*; VCH: New York, 1988; 361 spectra.

14. *Verification Database*, Vol. E1, EI Spectra of Chemical Warfare Agents, 1988; part of the *Finish Blue Books*, a 22-volume set relating to the Verification of the Chemical Weapons Convention; VERIFIN: Finish Institute of Helsinki, Finland (http://www.verifin.helsinki.fi).

15. Ardrey, RE; Allan, AE; Bal, TS; Joyce, JR; Moffat, AC *Pharmaceutical Mass Spectra*; The Pharmaceutical Press: London, 1985; 1065 spectra.

16. †Heller, SR; Milne, GWA; Gevantman, LH *EPA/NIH Mass Spectral Data Base, Supplement 2, 1983*; National Standard Reference Data System, National Bureau of Standards, Department of Commerce, United States Government, 1983; 6557 spectra.

17. †Sunshine I; Caplis, M *CRC Handbook of Mass Spectra of Drugs*; CRC: Boca Raton, FL, 1981; 1,208 EI spectra and 628 CI spectra.

18. †Middleditch, BS; Missler, SR; Hines, HB *Mass Spectrometry of Priority Pollutants*; Plenum: New York, 1981; 114 spectra.

19. †Heller, SR; Milne, GWA *EPA/NIH Mass Spectral Data Base, Supplement 1, 1980*; National Standard Reference Data System, National Bureau of Standards, Department of Commerce, United States Government, 1980; 8807 spectra.

20. †‡Heller, SR; Milne, GWA *EPA/NIH Mass Spectral Data Base*, 4 volumes and an index; National Standard Reference Data System, National Bureau of Standards, Department of Commerce, United States Government, 1978; 23,556 spectra.

21. †Cornu, A; Massot, R *Compilation of Mass Spectral Data*, 2nd ed., 2 volumes; Heyden: Philadelphia, PA, 1975; 10,000 spectra; 1st ed., 1966, 5000 spectra; 1st suppl., 1967, 1000 spectra; 2nd suppl., 1971, 1000 spectra.

22. Stenhagen, E; Abrahamsson, S; McLafferty, FW *Registry of Mass Spectral Data*, 4 volumes; Wiley: New York, 1974, ISBN:0471821152; 18,806 spectra.

23. Spiteller, M; Spiteller G *Massenspektrensammlung von Lösungmitteln, Verunreingungen, Säulenbelegmaterialien und einfachen aliphatischen Verbindungen*; Springer-Verlag: Wien-New York, 1973, ISBN:3211811176 hc; ISBN:0387811179 pbk.

24. †Safe, S; Hutzinger, O *Mass Spectrometry of Pesticides and Pollutants*; CRC: Cleveland, OH, 1973; 275 spectra.

25. Stenhagen, E; Abrahamsson, S; McLafferty, FW *Atlas of Mass Spectral Data*; Wiley: New York, 1969.

26. *Index of Mass Spectral Data*, AMD 11, 2nd ed.; American Society for Testing and Materials: Philadelphia, PA, 1969; 8000 spectra; 1st ed., 1963; 3200 spectra.

‡ Also available in electronic format
† Out of print

Mass Spectrometry, GC/MS, and LC/MS Journals

Articles containing information on mass spectrometry can be found in many different scientific journals as well as those listed below. This list consists of journals that are specific to mass spectrometry (Journals 1, 2, 3, 4, 5, 6, 10, 11, 15, 16) that pertain to a specific analytical technique (Journals 7 and 8) or that pertain to general chemistry (Journals 9, 12, 13, 14). Some of the journals have complementary subscriptions (Journals 12, 13, 14), whereas other journals have annual subscription rates of thousands of dollars (Journals 3, 4, 6, 7, 8). Some journals have reasonable society membership rates (Journals 1 and 9). The more expensive journals will often have reasonable individual subscription prices (Journals 3 and 4).

In addition to review and research articles, most of these journals also provide reviews of software, books, and other items of interest to the mass spectrometrist. The exceptions are the proceedings of meetings (Journals 15 and 16) and listings and/or abstract sources (Journals 10 and 11).

One of the interesting features of *JMS* (Journal 3) is a section entitled "Current Literature in Mass Spectrometry" that appears at the end of every issue. This feature is a bibliography of articles published over the past six to eight weeks. It is divided into 11 major sections with the Biology/Biochemistry section subdivided into 4 additional categories. At the end of each volume, all the listings for the year are made available in a Microsoft® Access format that can be searched electronically.

1. *Journal of the American Society for Mass Spectrometry*; Elsevier: New York.

2. *European Mass Spectrometry*; IM Publications: West Sussex, U.K.

3. *Journal of Mass Spectrometry*; Wiley: New York.
   (formerly *Organic Mass Spectrometry*, incorporating *Biomedical Mass Spectrometry*)

4. *Rapid Communications in Mass Spectrometry*; Wiley: New York.

5. *Mass Spectrometry Reviews*; Wiley: New York.

6. *International Journal of Mass Spectrometry*; Elsevier: New York.

7. *Journal of Chromatography, A*; Elsevier: New York.

8. *Journal of Chromatography, B*; Elsevier: New York.

9. *Analytical Chemistry*; American Chemical Society: Washington, DC.

10. *CA Selects Plus: Mass Spectrometry*; American Chemical Society: Washington, DC.

11. *Mass Spectrometry Bulletin*; Royal Society of Chemistry: Cambridge, U.K.

12. *American Laboratory*; ISC: Shelton, CT.

13. *LC/GC*; Advanstar: Eugene, OR.

14. *Spectroscopy*; Advanstar: Eugene, OR.

15. *Proceedings of the nth Annual ASMS Conference on Mass Spectrometry and Allied Topics* published annually; American Society for Mass Spectrometry: Santa Fe, NM (ASTM E14 Committee Meetings began in 1952; published annually from 1961–1969 ASTM E14 Committee Meeting Proceedings; ASMS began in 1970 with the 18th Conference Proceedings). Beginning with the 1999 meeting, the Proceedings are only available on CD-ROM.

16. *Advances in Mass Spectrometry*, Proceedings of the Triennial International Mass Spectrometry Conference. Vols. 1 (1958), 2 (1961), and 3 (Paris, France,1964) are found in the **Books of Historical Significance** section of this bibliography. Vols. 12 (Amsterdam, The Netherlands, 1991), 13 (Budapest, Hungary, 1994), and 14 (Tampere, Finland, 1997) are found in the **Reference Books** section of this bibliography. The 15th meeting was held in Barcelona, Spain, 2000. The 16th meeting was held in Edinburgh, United Kingdom, 2003. The 2006 meeting is scheduled for Prague, Czech Republic. The following is a complete listing of the series.

A   Volume 1 published by Pergamon: NY, London, Paris, LA; edited by JD Waldron, copyright 1959, London (1958).

B   Volume 2 published by Pergamon: Oxford, U.K.; MacMillan: New York; edited by RM Elliott, copyright 1963, Oxford (1961).

C   Volume 3 published by The Institute of Petroleum: London; edited by WL Mead, copyright 1966, Paris (1964).

D   Volume 4 published by The Institute of Petroleum: London; edited by E Kendrick, copyright 1967, Berlin (1967).

E   Volume 5 published by The Institute of Petroleum: London; edited by A Quayle, copyright 1971, Brussels (1970).

F   Volume 6 published by Elsevier: Amsterdam (Applied S); edited by The Institute of Petroleum (Staff) and AR West, copyright 1974, Edinburgh (1973).

G   Volume 7A and 7B published by Heyden & Son on behalf of The Institute of Petroleum: London; edited NR Daly, copyright 1978, Florence (1976).

H   Volume 8A and 8B published by Heyden & Son on behalf of The Institute of Petroleum: London; edited by A Quayle, copyright 1980, Oslo (1979).

I   Volume 9 published by Elsevier: Amsterdam; edited by ER Schmidt (JFK Huber, Chair); published in four volumes of the *International Journal of Mass Spectrometry and Ion Processes*, copyright Vol. 45, Dec. 1982; Vol. 46 and 47, Jan.1983; and Vol. 48, Feb. 1983, Vienna (1982).

J   Volumes 10A and 10B published by Wiley: Chichester, U.K.; edited by JFJ Todd, copyright 1986, Swansea (1985).

K   Volumes 11A and 11B published by Heyden & Son: London; edited by P Longevialle, copyright 1989, Bordeaux (1988).

L   Volume 12 published by Elsevier: Amsterdam; edited by PG Kistemaker and NM Nibbering, copyright 1992, Amsterdam (1991).

M   Volume 13 published by Wiley: Chichester, U.K.; edited by I Cornides, Gy Horváth, and K Vékey, copyright 1994, Budapest (1994).

N   Volume 14 published by Elsevier: Amsterdam; edited by EJ Karjalainen, AE Hesso, JE Jalonen, UP Karjalainen, copyright 1998 (CD also available), Tampere, Finland (1997).

O   Volume 15 published by Wiley: Chichester, U.K.; edited by E Gelpi, copyright 2001, Barcelona, Spain (2000).

P   Volume 16, published by Elsevier: Amsterdam; edited by AE Ashcroft, G Brenton, JJ Monaghan, copyright 2004, Edinburgh, U.K. (2003).

Inductively Coupled Plasma Mass Spectrometry

Mass spectrometers were used for the determination of inorganic ions before they found use in organic chemistry. A great deal of inorganic mass spectrometry now is associated with the use of inductively coupled plasma (ICP) sources. These ICP techniques had their origin in atomic emission spectroscopy but today are being dominated by mass spectrometry.

1. Nelms, S *Inductively Coupled Plasma Mass Spectrometry Handbook*; Taylor & Francis: London, 2005, ISBN:0849323819 hc; 356 pp.

2. Thomas, R *Practical Guide to ICP-MS*; Practical Spectroscopy Series, Vol. 33; Marcel Dekker: New York, 2003, ISBN:0824753194 hc; 324 pp.

3. De Laeter, JR *Applications of Inorganic Mass Spectrometry*; Wiley–Interscience Series on Mass Spectrometry; Desiderio, DM; Nibbering, NMM, Eds.; Wiley–Interscience: New York, 2001, ISBN:0471345393.

4. Barshick, CM; Duckworth, DC; Smith, DH *Inorganic Mass Spectrometry: Fundamentals and Applications*; Practical Spectroscopy Series, Vol. 24; Marcel Dekker: New York, 2000, ISBN:0824702433 (reviewed *JASMS* 11:822).

5. Beauchemin, D; Wood, TJ; Gregoire, DC; Gunther, D; Karanassions, V; Mermet, JM *Discrete Sample Introduction Technique for Inductively Coupled Plasma Mass Spectrometry*; Elsevier: New York, 2000, ISBN:0444899510.

6. Taylor, HE *Inductively Coupled Plasma-Mass Spectroscopy: Practices and Techniques*; Academic: San Diego, CA, 2000, ISBN:0126838658.

7. *Hill, SJ, Ed. *ICP Spectrometry and Its Applications*; Sheffield Academic: Sheffield, U.K., 1999, ISBN:0849397391 (reviewed *JASMS* 12:1226).

8. Montaser, A, Ed. *Inductively Coupled Plasma Mass Spectrometry*; VCH: Berlin, 1998, ISBN:0471186201.

9. Morrow, RW; Crain, JS, Eds. *Applications of Inductively Coupled Plasma-Mass Spectrometry to Radionuclide Determinations*, Vol. 2; American Society for Testing & Materials: Philadelphia, PA, 1998, ISBN:0803124961.

10. Abkar, M, Ed. *Inductively Coupled Plasma Mass Spectrometry: From A to Z*; VCH: Berlin, 1996, ISBN:1560819022.

11. Crain, JS; Morrow, RW, Eds. *Applications of Inductively Coupled Plasma-Mass Spectrometry to Radionuclide Determinations*; American Society for Testing & Materials: Philadelphia, PA, 1995, ISBN:0803120346.

12. Evans, EH *Inductively Coupled and Microwave Induced Plasma Sources for Mass Spectrometry*; Royal Society of Chemistry: Cambridge, U.K., 1995, ISBN:0854045600.

13. Jarvis, KE; Gray, AL; Houk, RS *Handbook of Inductively Coupled Plasma Mass Spectrometry*; Chapman Hall: London, 1991.

14. Date, AR *Applications of Inductively Coupled Plasma Mass Spectrometry*; Routledge: London, 1989, ISBN:0751401323.

15. Adams, F; Gijbels, R; van Grieken, R *Inorganic Mass Spectrometry*, Chemical Analysis: A Series of Monographs on Analytical Chemistry and Its Applications; Wiley: New York, 1988, ISBN:0471823643.

Personal Computer MS Abstract Sources

1. *Current Contents® on CD-ROM, Physical, Chemical & Earth Sciences*; Institute for Scientific Information: Philadelphia, PA; FREE Demo available.

2. *Analytical Abstracts on CD-ROM*; Royal Society of Chemistry: Cambridge, U.K.; or SilverPlatter Information: Norwood, MA; FREE 30-day Trial Subscription.

3. *CASurveyor: Mass Spectrometry and Applications*; Chemical Abstracts Service (American Chemical Society): Washington, DC; FREE Demo available.

4. *LC/MS Update* (1991–current); *GC/MS Update, Part A: Environmental* (1991–1996); *GC/MS Update, Part B: Biomedical, Clinical, Drugs* includes forensics (1991–current); HD Science: Newport, Wilmington, DE; or HD Science Limited: Nottingham, U.K.

5. *The PC Version of the Mass Spectrometry Bulletin*; Royal Society of Chemistry: Cambridge, U.K.

6. Annual Collections of the "Current Literature in Mass Spectrometry" in the *Journal of Mass Spectrometry* (1995–1998); Wiley: Chichester, U.K.

Integrated Spectral Interpretation Books

General integrated spectral interpretation books include information on the interpretation of proton NMR, IR, and mass spectra as well as how to use ultraviolet data in conjunction with these three spectral techniques. Some books also include a section on ^{13}C NMR. These books are good for an overview of the subject but do not provide the in-depth mass spectrometry interpretational information. In addition to the books listed below, the Mathieson book (Interpretation 26) is also an integrated book in that it includes separate sections on NMR and IR as well as the section on mass spectrometry.

1. Field, LD; Sternhell, S; Kalman, JR *Organic Structures from Spectra*, 2nd ed.; Wiley: Chichester, U.K., 2002, ISBN:0470843616; 1st ed., 1986; 2nd ed., 1995.

2. Crews, P; Jaspars, M; Rodriquez, J *Organic Structure Analysis*, 1st ed.; Oxford University Press: Oxford, U.K., 1998.

3. Lambert, JB; Shurvell, HF; Lightner, DA; Cooks, RG *Organic Structural Spectroscopy*; Prentice Hall: Upper Saddle River, NJ, 1998.

4. Silverstein, RM; Webster, FX *Spectrometric Identification of Organic Compounds*, 6th ed.; Wiley: New York, 1998; 1st ed., 1963; 2nd ed., 1967, Silverstein, RM; Bassler, GC; 3rd ed., 1974; 4th ed., 1981; 5th ed., 1991, Silverstein, RM; Bassler, GC; Morrill, TC.

5. Harwood, LM; Claridge, TDW *Introduction to Organic Spectroscopy*, 1st ed.; Oxford University Press: New York, 1997.

6. Hesse, M; Meier, H; Zeeh, B *Spectroscopic Methods in Organic Chemistry*, 1st ed.; Linden, A; Murray, M, Translators; Thieme: New York, 1997.

7. Pavia, DL; Lampman, GM; Kriz, GS *Introduction to Spectroscopy: A Guide for Students of Organic Chemistry*, 2nd ed.; Saunders College: Orlando, FL, 1996, ISBN:0030584272; 1st ed., 1979, ISBN:0721671195.

8. Feinstein, K *Guide to Spectroscopic Identification of Organic Compounds*; CRC: Boca Raton, FL, 1995.

9. Williams, DH; Fleming, I *Spectroscopic Methods in Organic Chemistry*, 5th ed.; McGraw-Hill: London, 1995; 1st ed., 1966.

10. Jones, C; Mulloy, B; Thomas, AH, Eds. *Spectroscopic Methods and Analyses (NMR, Mass Spectrometry, and Metalloprotein Techniques)*, Vol. 17 *Methods in Molecular Biology*; Humana: Totowa, NJ, 1993.

11. Kemp, W *Organic Spectroscopy*, 3rd ed.; W. H. Freeman: New York, 1991.

12. Fresenius, W; Huber, JFK; Pungor, E; Rechnitz, GA; Simon, W; West, TS, Eds. *Tables of Spectral Data for Structure Determination of Organic Compounds*, 2nd English ed. (translated from the German edition by K. Biemann); Springer-Verlag: Berlin, Germany, 1989.

13. Sorrell, TN *Interpreting Spectra of Organic Molecules*; University Science: Mill Valley, CA, 1988.

14. Lambert, JB; Shurvell, HF; Lightner, DA; Cooks, RG *Introduction to Organic Spectroscopy*; MacMillan: New York, 1987, ISBN:002363001.

15. Lambert, JB; Shurvell, HF; Verbit, L; Cooks, RG; Stout, GH *Organic Structural Analysis*; MacMillan: New York, 1978, ISBN:0023672900.

16. Scheinmann, F, Ed. *An Introduction to Spectroscopic Methods for the Identification of Organic Compounds*, Vol. 2 *Mass Spectrometry, Ultraviolet Spectroscopy, Electron Spin Resonance Spectroscopy, NMR (Recent Developments), Use of Various Spectral Methods Together, and Documentation of Molecular Spectra*; Pergamon: Oxford, U.K., 1974.

17. Mathieson, DW, Ed. *Interpretation of Organic Spectra*; Academic: New York, 1965.

Monographs

All the citations in this segment are from VG Instruments/Micromass. All instrument manufacturers publish application notes; however, Micromass (and its preceding companies) is the only manufacturer that has published this type of general-topic monograph. These monographs are like review articles found in *Mass Spectrometry Reviews* or the Special Features section of the *Journal of Mass Spectrometry*. These monographs are not as well referenced as the articles in these two journals but do provide a good overview of the subject. Such promotional material is of benefit to those wanting to get a quick understanding of a topic, and it is hoped that more of this type of material will be forthcoming.

Another example of the ready-reference-material approach to information dissemination is found in the Siuzdak book (Introductory 12). This book was written to provide a quick understanding of mass spectrometry to biotechnology executives who have to make financial decisions about mass spectrometry instrumentation and facilities.

1. Rose, ME *Modern Practice of Gas Chromatography/Mass Spectrometry*, VG Monographs in Mass Spectrometry, No. 1; VG Instruments: Altrincham, U.K.

2. Mellon, FA *Liquid Chromatography/Mass Spectrometry*, VG Monographs in Mass Spectrometry, No. 2; VG Instruments: Altrincham, U.K.

3. Clench, MR *A Comparison of Thermospray, Plasmaspray, Electrospray and Dynamic FAB*, VG Monographs in Mass Spectrometry, No. 3; VG Instruments: Altrincham, U.K.

4. Scrivens, JH; Rollins, K *Tandem Mass Spectrometry*, VG Monographs in Mass Spectrometry, No. 4; VG Analytical, Fisons Instruments: Altrincham, U.K.

5. Ardrey, B *Mass Spectrometry in the Forensic Sciences*, VG Monographs in Mass Spectrometry, No. 5; VG Analytical, Fisons Instruments: Altrincham, U.K.

6. Hsu, J-P *High and Low Resolution GC/MS in Environmental Sciences*, VG Monographs in Mass Spectrometry, No. 6; VG Analytical, Fisons Instruments: Altrincham, U.K.

7. Micromass *Back to Basics*, CD01 Version 2, electronic document from Dec. 2000; Manchester, U.K. (http://www.micromass.co.uk).

Software

There are several programs that are available as self-training. Those programs developed by the United Kingdom company, Cognitive Solutions (Software 12–15), are somewhat like English roast beef. They are intellectually nutritious; however, they fail to excite the experiential palate. Equivalent titles available from Academy Savant (Software 16–27) will hold the user's interest in a much more conducive manner for learning. Just as the *MS Fundamentals* program is a seminal tool in the development of the understanding of the transmission quadrupole mass spectrometer, *Fundamentals of GC/MS* is one of the better instrument-user software packages developed. This program was initially developed and sent out several times for review to a number of people who are involved in training on various aspects of GC/MS and in the development of training programs. The result is what is assured to become an award-winning effort.

All of the training programs from Academy Savant and Cognitive Solutions were developed in Interactive ToolBook, a powerful tool for the development of training programs. All have tests built into the programs to allow the user to evaluate the results of the training.

The two volumes of *SpectraBook* are also based on the Interactive ToolBook platform. These two programs each contain data on 50 separate compounds: mass, proton and ^{13}C NMR, and infrared spectra as well as the structure; molecular mass based on the atomic mass of each compound's elements; physical properties; and several synonyms. Help files are provided to assist the user in developing desired interpretational skills. Another nice feature is the ability to display which properties result in specific spectral peaks. Placing the Mouse pointed on a labeled mass spectral peak and holding down the left Mouse button will result in a display of the mechanism(s) that produced the ion represented by that peak. Similar displays are provided for the other types of spectra.

It is unfortunate that the author of *SpectraBook* (programs copyrighted in 1990 and 1992) did not take more care to be correct in some of the presentations such as the use of *m/z* as the symbol for mass-to-charge (inappropriately written Mass to Charge on the abscissa of mass spectra) ratio instead of the m/e symbol, which was replaced in the 1970s. The indicated shift of pairs of electrons in the displayed mechanism for β cleavage resulting from a γ-hydrogen shift does not instill confidence in the accuracy of instruction.

The programs published by ChemSW (Software 1–11) are very well thought out and provide utilities that are not found in the data system software for most, if not all, commercially available instruments. The titles of these programs are self-explanatory.

The two programs associated with the Wiley and NIST Mass Spectral Databases (Software 36 and 38) are widely available from a number of different sources. A number of GC/MS and LC/MS programs now provide the NIST Mass Spectral Search Program as the search routine used with their proprietary instrument software. Both of the programs are capable of reading most, if not all, commercially available instrument data formats.

1. *CESAR*™ Capillary Electrophoresis Simulation for Application Research; ChemSW: Fairfield, CA.

2. *GC-SOS*™ Gas Chromatography Simulation and Operation Software, Ver. 5; ChemSW: Fairfield, CA.

3. Cody, RB *MS Tools*; ChemSW: Fairfield, CA.

4. Michael, D *ChemSite Pro*; ChemSW: Fairfield, CA.

5. Bernert, JT, Jr. (Quadtech Associates) *Mass Spec Calculator*™ Ver. 3, and *Mass Spec Calculator*™ *Pro*; ChemSW: Fairfield, CA.

6. Junk, T *GC and GC/MS File Translator*™ *Professional*; ChemSW: Fairfield, CA.

7. Junk, T *GC and GC/MS File Manager*™; ChemSW: Fairfield, CA.

8. Bernert, JT, Jr. (Quadtech Associates) *Mass Differential Analysis Tools*; ChemSW: Fairfield, CA.

9. *Protein Tools*™; ChemSW: Fairfield, CA.

10. *HPLC Optimization*™; ChemSW: Fairfield, CA.

11. *GPMAW*™ General Protein Mass Analysis for Windows; ChemSW: Fairfield, CA.

12. *Interactive Training Program™ *Gas Chromatography*; published by Cognitive Solutions: Glasgow, U.K., a.k.a. Softbooks out of the United States; distributed by ChemSW: Fairfield, CA.

13. ‡Interactive Training Program™ *High Performance Liquid Chromatography*; published by Cognitive Solutions: Glasgow, U.K., a.k.a. Softbooks out of the United States; distributed by ChemSW: Fairfield, CA.

14. ‡Interactive Training Program™ *Advanced Gas Chromatography*; published by Cognitive Solutions: Glasgow, U.K., a.k.a. Softbooks out of the United States; distributed by ChemSW: Fairfield, CA.

15. ‡Davis, S (HD Technologies) Interactive Training Program™ *Mass Spectrometry*; published by Cognitive Solutions: Glasgow, U.K., a.k.a. Softbooks out of the United States; distributed by ChemSW: Fairfield, CA.

16. ‡Saunders, D *Introduction to Gas Chromatography*; Academy Savant[†]: Fullerton, CA.

17. ‡Saunders, D *Fundamentals of Gas Chromatography/Mass Spectrometry*; Academy Savant: Fullerton, CA.

18. ‡Saunders, D *Introduction to LC/MS*; Academy Savant: Fullerton, CA.

19. ‡*Introduction to High Performance Liquid Chromatography*; Academy Savant: Fullerton, CA.

20. ‡*Method Development in High Performance Liquid Chromatography*; Academy Savant: Fullerton, CA.

21. ‡*High Performance Liquid Chromatography Equipment*; Academy Savant: Fullerton, CA.

22. ‡*Troubleshooting High Performance Liquid Chromatography*; Academy Savant: Fullerton, CA.

23. ‡*Separation Modes of High Performance Liquid Chromatography*; Academy Savant: Fullerton, CA.

24. ‡*HPLC Calculation Assistant & Reference Tables*; Academy Savant: Fullerton, CA.

25. ‡*Identification & Quantification for HPLC*; Academy Savant: Fullerton, CA.

26. ‡Hart, M *MS Fundamentals*; published by Hewlett-Packard: Palo Alto, CA; distributed by Academy Savant: Fullerton, CA.

27. ‡Schatz, PF *SpectraBook*, Vol. 1, 1990 and Vol. 2, 1992; published by Falcon Software; distributed by Academy Savant: Fullerton, CA.

28. *DryLab*; LC Resources: Lafayette, CA.

29. *IST for GC/MS*; Ion Signature Technology, Inc.; North Smithfield, RI.

30. *Introduction to CE*; LC Resources: Lafayette, CA.

31. Figueras, J *Mass Spec*, Ver. 3.0; Trinity Software: Plymouth, NH.

32. *ACD/SpecManager: MS Module*; Advanced Chemistry Development: Toronto, Canada.

33. *ACD/IntelliXtract*; Advanced Chemistry Development: Toronoto, Canada.

34. *Mass Frontier*; ThemoQuest/HighChem: San Jose, CA.

35. *MassWorks*™; Cerno Bioscience: Danbury, CT.

36. *MASSTransit*; Palisade: Newfield, NY.

37. Antolasic, F *Wsearch: Mass Spectral Search Program*, last updated 1 October 1999. FreeWare (http://minyos.its.rmit.edu.au/~rcmfa/)

38. *Benchtop PBM* with *Wiley Registry of Mass Spectral Data*, Ver. 7 or 7N, or Select; Palisade: Newfield, NY.

39. *Wiley Registry of Mass Spectral Data*, Ver. 8 with structures; John Wiley & Sons: Hoboken, NJ.

40. Dahl, D *SIMION 3D*, Ver. 6.0, Ion and Electron Optics Program; Scientific Instrument Services: Ringoes, NJ.

41. *NIST Mass Spectral Search Program for Windows*, Ver. 2.0 and *NIST/EPA/NIH Mass Spectral Database* (*NIST05*) with *AMDIS*, Ver. 2.1; Automated Mass Spectral Deconvolution and Identification System, National Standard Reference Data System, National Institute of Standards and Technology: Gaithersburg, MD.

‡ Uses Asymetrix ToolBook Runtime
† Sloane Audio Visuals for Analysis and Training

WEB SITES FOR SOFTWARE COMPANIES:

ChemSW	http://www.chemsw.com
Cerno Bioscience	http://www.cernobioscience.com/
Ion Signature	http://www.ionsigtech.com/
LC Resources	http://www.lcresources.com
NIST	http://www.nist.gov/srd/analy.htm
Palisade	http://www.palisade.com
Scientific Instrument Services	http://www.sisweb.com
John Wiley & Sons	http://www.wiley.com/go/databases
Trinity Software	http://www.trinitysoftware.com
Advanced Chemistry Development	http://www.acdlabs.com
HighChem	http://www.highchem.com
Academy Savant	http://www.academysavant.com

INDEX

A

Symbols for LC/MS
Picograms–Scan Mode

MS Scan Modes

+ ESI	▬	Positive ESI, using a Q or a QqQ instrument; <u>regular scan mode</u>.
+ APCI	762	Positive APCI, using a Q or a QqQ instrument; <u>SIM</u> of m/z 762.
EI	▬	Particle Beam or Moving Belt with an EI source, using a Q or a QqQ instrument; <u>regular scan mode</u>.
+ ESI	⌐⌐	Positive ESI, using a Q or a QqQ instrument; <u>SIM</u> of more than one m/z value.
+ FAB	⬠ ≈	Positive CF-FAB, using a BE sector instrument; <u>regular scan mode</u>.
⌐⌐ + CI	⬠(28) ≈	Particle Beam or Moving Belt with a positive CI source, using a BE sector instrument, electric sector (E) <u>SIM,</u> and magnetic sector (B) set at m/z 281 at full acceleration voltage.
+ ESI ✷ 40	▬	Positive ESI, using a Q or a QqQ instrument; <u>skimmer-CID</u> (sCID) at 40 V offset and <u>regular scan mode</u>.
-/+ ESI	-/+ ⌐⌐	Dual polarity ESI programmed sCID, using a Q or a QqQ instrument; fragment ion detection in <u>SIM</u> mode in low-mass region; positive ESI for molecular ion detection in high-mass region, using <u>regular scan function</u>. Polarity switch occurs when sCID offset is switched off.
-/+ ESI	-/+/+ ▬	Dual polarity ESI programmed sCID, using a Q or a QqQ instrument; negative and positive polarity for fragment ion detection in <u>regular scan mode</u>; positive polarity for molecular ion detection in high-mass region, using <u>regular scan mode</u>. Polarity switch occurs when sCID offset is switched off.

Adapted with permission from Lehmann, WD "Pictograms for Experimental Parameters in Mass Spectrometry" *J. Am. Soc. Mass Spectrom.* **1997**, *8, 756–759.*

Pictograms–MS/MS Modes

MS Scan Modes

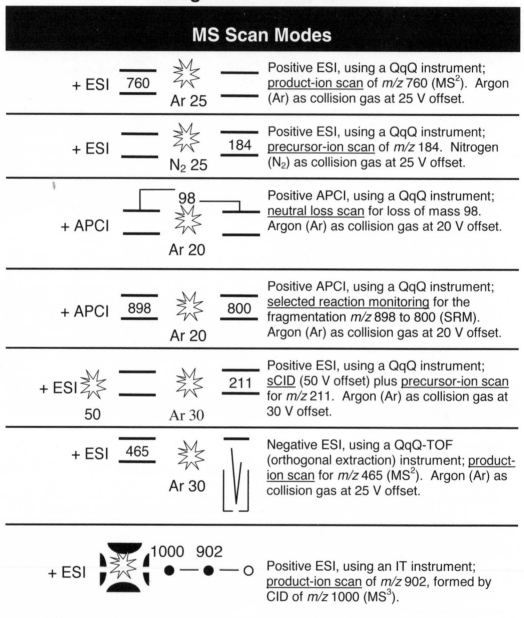

+ ESI ⎯ 760 ☀ ⎯ Ar 25

Positive ESI, using a QqQ instrument; <u>product-ion scan</u> of *m/z* 760 (MS^2). Argon (Ar) as collision gas at 25 V offset.

+ ESI ⎯ ☀ 184 N$_2$ 25

Positive ESI, using a QqQ instrument; <u>precursor-ion scan</u> of *m/z* 184. Nitrogen (N$_2$) as collision gas at 25 V offset.

+ APCI 98 ☀ Ar 20

Positive APCI, using a QqQ instrument; <u>neutral loss scan</u> for loss of mass 98. Argon (Ar) as collision gas at 20 V offset.

+ APCI 898 ☀ 800 Ar 20

Positive APCI, using a QqQ instrument; <u>selected reaction monitoring</u> for the fragmentation *m/z* 898 to 800 (SRM). Argon (Ar) as collision gas at 20 V offset.

+ ESI ☀ 50 ☀ Ar 30 211

Positive ESI, using a QqQ instrument; <u>sCID</u> (50 V offset) plus <u>precursor-ion scan</u> for *m/z* 211. Argon (Ar) as collision gas at 30 V offset.

+ ESI 465 ☀ Ar 30

Negative ESI, using a QqQ-TOF (orthogonal extraction) instrument; <u>product-ion scan</u> for *m/z* 465 (MS^2). Argon (Ar) as collision gas at 25 V offset.

+ ESI 1000 902

Positive ESI, using an IT instrument; <u>product-ion scan</u> of *m/z* 902, formed by CID of *m/z* 1000 (MS^3).

Adapted with permission from Lehmann, WD "Pictograms for Experimental Parameters in Mass Spectrometry" *J. Am. Soc. Mass Spectrom.* **1997**, *8*, 756–759.

Proof of a Molecular Ion Peak – M$^{+\bullet}$

1. If a compound is known, the molecular ion has a mass-to-charge ratio (*m/z*) value equal to the sum of the atomic masses of the most abundant isotope of each element that comprises the molecule (assuming the ion is a single-charge ion).

2. The nominal mass of a compound, or the *m/z* value for the molecular ion, is an even number for any compound containing only C, H, O, S, Si, P, and the halogens.

    Fragment ions, derived via homolytic, heterolytic, or sigma-bond cleavage from these molecular ions (even *m/z*), have an odd *m/z* value and an even number of electrons.

    Fragment ions derived from these molecular ions (even *m/z*) via expulsion of neutral components (e.g., H_2O, CO, ethylene, etc.) have an even *m/z* value and an odd number of electrons.

3. **Nitrogen rule**: A compound containing an odd number of nitrogen atoms—in addition to C, H, O, S, Si, P, and the halogens—has an odd nominal mass.

    Molecular ions of these compounds fragment via homolytic, heterolytic, or sigma-bond cleavage to produce ions of an even *m/z* value unless the nitrogen atom is lost with the neutral radical.

    An even number of nitrogen atoms in a compound results in an even nominal mass.

4. The molecular ion peak must be the highest *m/z* value of any significant (nonisotope or nonbackground) peak in the spectrum. Corollary: The highest *m/z* value peak observed in the mass spectrum need not represent a molecular ion.

5. The peak at the next lowest *m/z* value in the mass spectrum must not correspond to the loss of an impossible or improbable combination of atoms.

6. No fragment ion may contain a larger number of atoms of any particular element than the molecular ion.

Courses on the interpretation of mass spectra and techniques of mass spectrometry are offered by the Continuing Education Department of the American Chemical Society (http://www.acs.org/shortcourses).

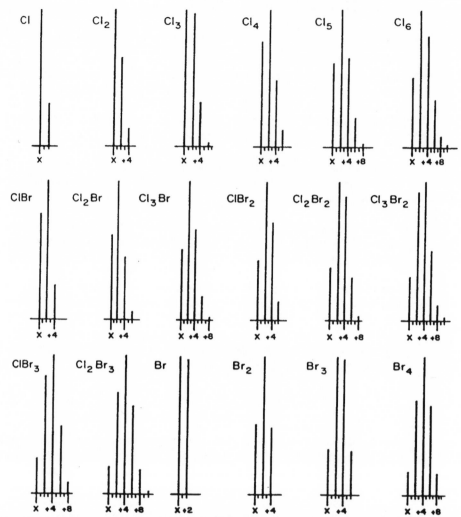

Graphical representation of relative isotope peak intensities for any given ion containing the indicated number of chlorine and/or bromine atoms. Numeric values are on the opposite page.

Steps to Determine Elemental Composition Based on Isotope Peak Ratios

1. Determine the nominal m/z value peak (peak at lowest m/z value, above which other peaks can be attributed to isotopic multiplicity or background).
2. Assign the X+2 elements, except oxygen.
3. Assign the X+1 elements (remember to normalize X+1 to X, if necessary).
4. Balance the mass.
5. Assign the atoms of oxygen.
6. Balance the mass.
7. Assign the X elements.
8. From the elemental composition, determine the number of rings plus double bonds.
9. Propose a possible structure.
10. Does it make sense?

Atoms of ClBr	X	X+2	X+4	X+6	X+8	X+10
Cl	100	32.5				
Cl$_2$	100	65.0	10.6			
Cl$_3$	100	97.5	31.7	3.4		
Cl$_4$	76.9	100	48.7	0.5	0.9	
Cl$_5$	61.5	100	65.0	21.1	3.4	0.2
Cl$_6$	51.2	100	81.2	35.2	8.5	1.1
ClBr	76.6	100	24.4			
Cl$_2$Br	61.4	100	45.6	6.6		
Cl$_3$Br	51.2	100	65.0	17.6	1.7	
ClBr$_2$	43.8	100	69.9	13.7		
Cl$_2$Br$_2$	38.3	100	89.7	31.9	3.9	
Cl$_3$Br$_2$	31.3	92.0	100	49.9	11.6	1.0
ClBr$_3$	26.1	85.1	100	48.9	8.0	
Cl$_2$Br$_3$	20.4	73.3	100	63.8	18.7	2.0
Br	100	98.0				
Br$_2$	51.0	100	49.0			
Br$_3$	34.0	100	98.0	32.0		
Br$_4$	17.4	68.0	100	65.3	16.0	

Chlorine and Bromine Isotopic Abundance Ratios

M – 1	loss of hydrogen radical	M – $^{\bullet}$H
M – 15	loss of methyl radical	M – $^{\bullet}$CH$_3$
M – 29	loss of ethyl radical	M – $^{\bullet}$CH$_2$CH$_3$
M – 31	loss of methoxyl radical	M – $^{\bullet}$OCH$_3$
M – 43	loss of propyl	M – $^{\bullet}$CH$_2$CH$_2$CH$_3$
M – 45	loss of ethoxyl	M – $^{\bullet}$OCH$_2$CH$_3$
M – 57	loss of butyl radical	M – $^{\bullet}$CH$_2$CH$_2$CH$_2$CH$_3$
M – 2	loss of hydrogen	M – H$_2$
M – 18	loss of water	M – H$_2$O
M – 28	loss of CO or ethylene	M – CO or M – C$_2$H$_4$
M – 32	loss of methanol	M – CH$_3$OH
M – 44	loss of CO$_2$	M – CO$_2$
M – 60	loss of acetic acid	M – CH$_3$CO$_2$H
M – 90	loss of silanol: HO-Si(CH$_3$)$_3$	M – HO-Si-(CH$_3$)$_3$

Common Neutral Losses

Amino Acids with Apolar Side Chains

Name And Res Comp	Abb	Res Nom	Residue Monoiso	Residue Ave.	Immo Mass	SC Mass	Structure
Glycine C_2H_3NO	gly G	57	57.02146	57.0520	30	–	
Alanine C_3H_5NO	ala A	71	71.03711	71.0788	44	15	
Valine C_5H_9NO	val V	99	99.06841	99.1326	72	43	
Leucine $C_6H_{11}NO$	leu L	113	113.08406	113.1595	86	57	
Isoleucine $C_6H_{11}NO$	ile I	113	113.08406	111.1595	86	57	
Proline C_5H_7NO	pro P	97	97.05276	97.1167	70	–	
Phenyl-alanine C_9H_9NO	phe F	147	147.06841	147.1766	120	91	
Tryptophan $C_{11}H_{10}N_2O$	trp W	186	186.07931	186.2133	159	130	
Methionine C_5H_9NOS	met M	131	131.04049	131.1986	104	75	

Amino Acids with Uncharged Polar Side Chains

Name And Res Comp	Abb	Res Nom	Residue Monoiso	Residue Ave.	Immo Mass	SC Mass	Structure
Serine $C_3H_5NO_2$	ser S	87	87.03203	87.0782	60	31	
Threonine $C_4H_7NO_2$	thr T	101	101.04768	101.1051	74	45	
Cysteine C_3H_5NOS	cyc C	103	103.00919	103.1448	76	47	
Tyrosine $C_9H_9NO_2$	tyr Y	163	163.06333	163.1760	136	107	
Asparagine $C_4H_6N_2O_2$	asn N	114	114.04293	114.1039	87	58	
Glutamine $C_5H_8N_2O_2$	gln Q	128	128.05856	128.1308	101	72	

Amino Acids with Charged Polar Side Chains

Name And Res Comp	Abb	Res Nom	Residue Monoiso	Residue Ave.	Immo Mass	SC Mass	Structure
Aspartic acid $C_4H_5NO_3$	asp D	115	115.02694	115.0886	88	59	
Glutamic acid $C_5H_7NO_3$	glu E	129	129.04259	129.1155	102	73	
Lysine $C_6H_{12}N_2O$	lys K	128	128.09496	128.1742	101	72	
Arginine $C_6H_{12}N_4O$	arg R	156	156.10111	156.1876	129	100	
Histidine $C_6H_7N_3O$	his H	137	137.05891	137.1412	110	81	

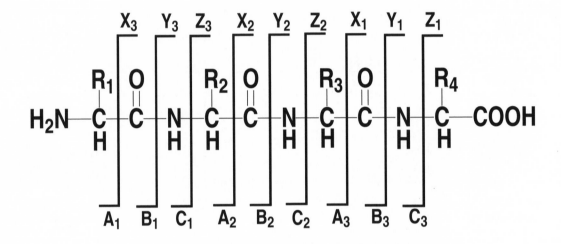

The general fragmentation of a generic-protonated peptide
according to the nomenclature of Peter Roepstorff

Amino acid residue masses

Ion types found where the amino acid content of the ion differs by one residue

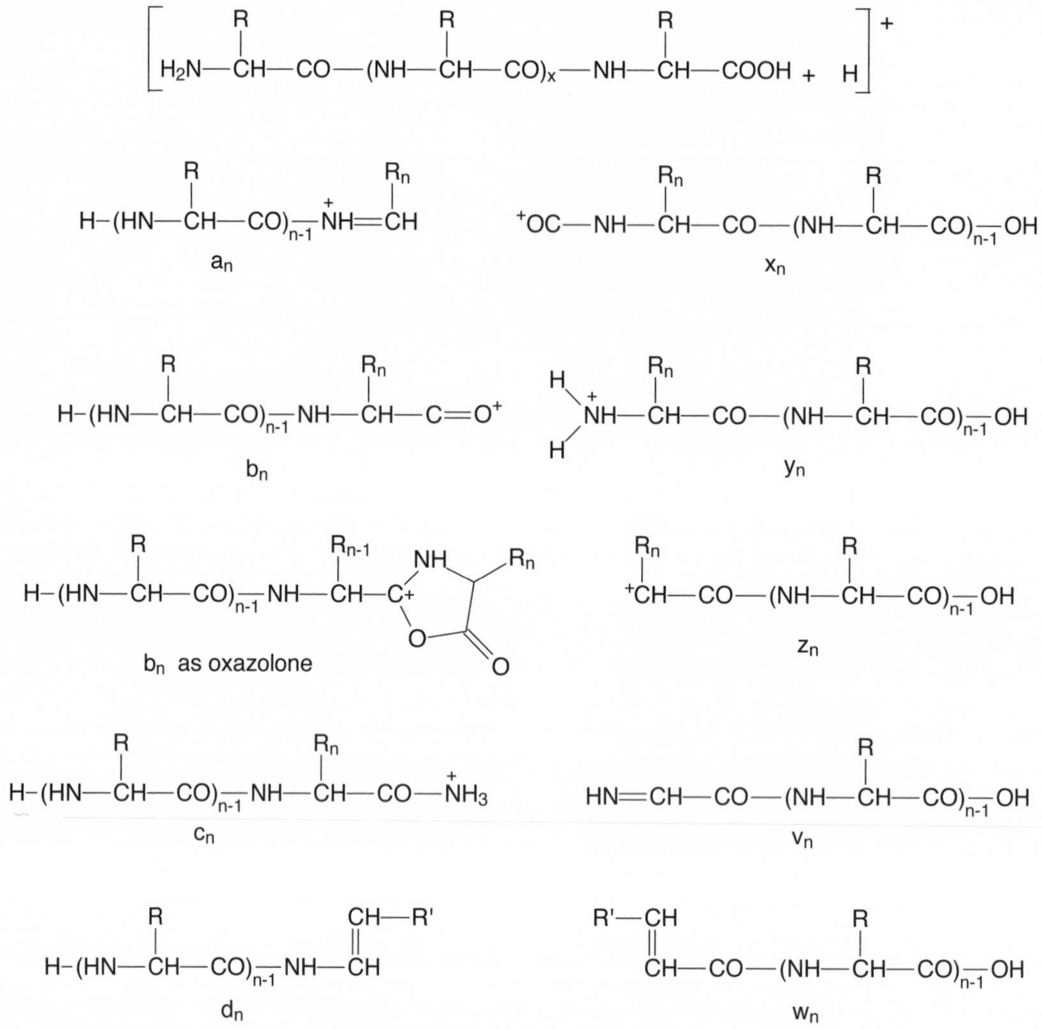

$$\left[H_2N-\underset{\underset{R}{|}}{CH}-CO-(NH-\underset{\underset{R}{|}}{CH}-CO)_x-NH-\underset{\underset{R}{|}}{CH}-COOH + H \right]^+$$

$$H-(HN-\underset{\underset{R}{|}}{CH}-CO)_{n-1}\overset{+}{N}H=\underset{\overset{|}{R_n}}{CH}$$

a_n

$$^+OC-NH-\underset{\overset{|}{R_n}}{CH}-CO-(NH-\underset{\underset{R}{|}}{CH}-CO)_{n-1}-OH$$

x_n

$$H-(HN-\underset{\underset{R}{|}}{CH}-CO)_{n-1}-NH-\underset{\overset{|}{R_n}}{CH}-C{=}O^+$$

b_n

$$\underset{H}{\overset{H}{\diagdown}}\overset{+}{N}H-\underset{\overset{|}{R_n}}{CH}-CO-(NH-\underset{\underset{R}{|}}{CH}-CO)_{n-1}-OH$$

y_n

$$H-(HN-\underset{\underset{R}{|}}{CH}-CO)_{n-1}-NH-\underset{\overset{|}{R_{n-1}}}{CH}-C^+$$

b_n as oxazolone

$$^+CH-CO-(NH-\underset{\underset{R}{|}}{CH}-CO)_{n-1}-OH$$
$\underset{R_n}{}$

z_n

$$H-(HN-\underset{\underset{R}{|}}{CH}-CO)_{n-1}-NH-\underset{\overset{|}{R_n}}{CH}-CO-\overset{+}{N}H_3$$

c_n

$$HN=CH-CO-(NH-\underset{\underset{R}{|}}{CH}-CO)_{n-1}-OH$$

v_n

$$H-(HN-\underset{\underset{R}{|}}{CH}-CO)_{n-1}-NH-\underset{\overset{||}{CH-R'}}{CH}$$

d_n

$$\underset{\overset{||}{CH}}{R'-CH}-CO-(NH-\underset{\underset{R}{|}}{CH}-CO)_{n-1}-OH$$

w_n

Specific structures for the a, b, c and x, y, z ions per the Klaus Biemann nomenclature

To recognize the start of a **y** series:

    a) it will be a high-mass ion

    b) it will correspond to the loss of a residue mass from MH^+

    c) note the **y** ion accumulates two hydrogens:
          1) one from the original protonation
          2) the other is abstracted from the N-terminal portion as it is expelled

To recognize the start of a **b** series:

    a) it will be a high-mass ion

    b) it will correspond to the loss of a residue plus H_2O from MH^+

b_n as oxazolone

Practical Considerations for LC/MS

Common Mobile-Phase Clusters in Positive-Ion Mode

Acetonitrile		Methanol		Water	
42	$(CH_3CN)H^+$	33	$(CH_3OH)H^+$	19	$(H_2O)H^+$
83	$(CH_3CN)_2H^+$	65	$(CH_3OH)_2H^+$	37	$(H_2O)_2H^+$
124	$(CH_3CN)_3H^+$	97	$(CH_3OH)_3H^+$	55	$(H_2O)_3H^+$
165	$(CH_3CN)_4H^+$	129	$(CH_3OH)_4H^+$	73	$(H_2O)_4H^+$
206	$(CH_3CN)_5H^+$	161	$(CH_3OH)_5H^+$	91	$(H_2O)_5H^+$
247	$(CH_3CN)_6H^+$			199	$(H_2O)_{11}H^+$
288	$(CH3CN)_7H^+$			379	$(H_2O)_{21}H^+$
				505	$(H_2O)_{28}H^+$

Common Adducts

Positive Ionization Mode

M + 23 (Na)
M + 32 (MeOH)
M + 39 (K)
M + 41 (CH$_3$CN)
m/z 159 (TFA + Na)
m/z 242 (tetrabutyl ammonium)
m/z 391 (DOP)

Negative Ionization Mode

M + 45 (Formate)
M + 60 (Acetate)
M + 58 (NaCl salt)
M + 78 (DMSO)
M + 113 (TFA)

Common Artifacts in Negative-Ion Mode

26   CN^{1-} – from acetonitrile
35   Cl^{1-} – from inorganic or organic chlorides
59   Acetate^{1-}
79   Phosphate PO_3^- (several sources: phosphoric acid, oligonucleotides)
80   Sulfate SO_3^-
96   SO_4^{2-} adduct (proteins and peptides)
97   HSO_4^- and $H_2PO_4^-$
113   TFA^{1-}

Common Artifacts and Adducts in Positive-Ion Mode

28	Series of peaks m/z 300 to 600 separated by 28 m/z units – triglycerides from fingerprints or contamination
41	Acetonitrile adduct
44	Series of peaks separated by 44 m/z units – ethylene oxide
50	Series of peaks separated by 50 m/z units – CF2 fluorinated surfactants
58	Series of peaks separated by 58 m/z – propylene oxide, or NaCl adduct if 58/60
61.5	Series of peaks separated by 61.5 m/z – copper adducts
63	$H+H_2CO_3$
64	ACN+Na
71/72	THF
74	DMF + H or diethyl amine + H^+
79	DMSO + H^+
83 2	ACN + H^+
88	Formic acid + acetonitrile + H^+
101, 138, 183	MeOH + H_2O clusters
102	Triethylamine + H^+
102	AcN + HOAc
105	2× acetonitrile + Na^+
146	3× acetonitrile + Na^+
149	Fragment from dioctyl phthalate
158	Amino sugar; common from antibiotic analyses
159	NaTFA + Na^+
163	Nicotine
169, 165, 195	(possibly 133 and 135) – dimethyl phthalate
181	BHA (butylated hydroxyanisole – food additive, preservative)
186	Tributylamine + H^+ (ion-pair reagents)
211, 227, 241, 253, 269, 281	Detergents from glassware
219	Tri-*tert*-butylphosphine oxide from peptide synthesis
221	BHT (butylated hydroxytoluene – food additive, preservative)
241, 253, 255, 269, 281	Fatty acids and/or soap
279	Dibutyl phthalate
281/282	Oleic acid soap or oleamide (Na^+, K^+ or NH_4^+ from mold release agents used in plastic production)
317, 361, 405	Triton detergent
362	Dioctyl diphenyl phthalate
388, 437, 444, 463	Lubricants from HPLC components
391	Dioctyl phthalate + H^+ (common from contaminated solvents and from plastic tubing)
413	Dioctyl phthalate + Na^+
427	Dioctyl sebacate
447	Diisodecyl phthalate + H^+
481, 525, 569	PEG as sodium adducts
503, 547, 591	protonated PEG
563	Oleic acid soap

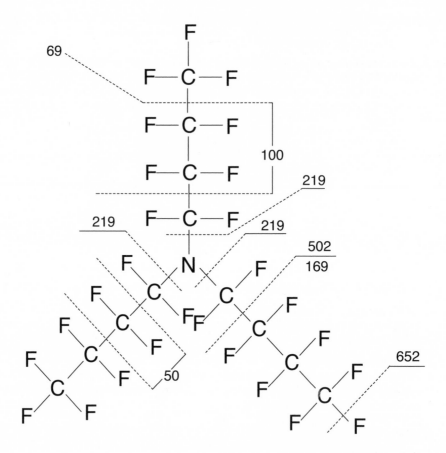

The molecular mass of perfluorotributylamine (PFTBA a.k.a. FC-43), used to calibrate the *m/z* scale of transmission quadrupole and quadrupole ion trap mass spectrometers operated in the electron ionization mode, is 671. The following is an explanation of the origin of some of the peaks observed in its EI mass spectrum:

671			671		671	
−207	(3 × 69)		−507	(3 × 169)	−57	(3 × 19)
464			164		614	
−50			+100			
414			264		169	
					−38	(2 × 19)
					131	

Other Books from Global View Publishing

Introduction to Mass Spectrometry, 4th edition

J. Throck Watson and O. David Sparkman

This quintessential text on mass spectrometry has been the most widely used mass spectrometry textbook since the first edition appeared in 1976. The fourth edition is the only introductory text on mass spectrometry that covers all topics related to organic and biological molecules as well as all aspects of the instrumentation. This book gives details on EI, MALDI, electrospray (ESI), GC/MS, LC/MS, all types of m/z analyzers, data interpretation of mass spectra regardless of their origins, and much more. Included are thousands of citations. All cited journal articles include titles and page-number ranges, which make this book the definitive resource of all areas of mass spectrometry. This book also exhibits the teaching skills developed by these 2 widely acclaimed authors over the past 30 years.

First edition, 1976	ISBN: 0-89004-056-7
Second edition, 1985	ISBN: 0-88167-081-2
Third edition, 1997	ISBN: 0-397-51688-6
Fourth edition, 2006	ISBN: 0-9660813-7-4

A Global View of LC/MS: How to solve your most challenging analytical problems

Ross Willoughby, Ed Sheehan, and Sam Mitrovich

This text is a definitive problem-solving guide to the rapidly expanding field of liquid chromatography/mass spectrometry. A structured approach is utilized throughout the text to evaluate and solve many of the problems faced in the analysis of complex and labile samples. The reader is presented with techniques to evaluate and select the many alternative technologies that are available today with LC/MS. You will learn to:

Evaluate the various LC/MS technologies	Maintain an effective LC/MS laboratory
Evaluate your needs for LC/MS	Solve real analytical problems
Acquire LC/MS	Develop an individualized plan
Set up an effective LC/MS laboratory	Develop and validate LC/MS methods

First edition, 1998	ISBN: 0-9660813-0-7
Second edition, 2001	ISBN: 0-9660813-5-8

LC/MS Methods: An essential tool to accelerate methods development

Stephen Down and John Halket, Editors

This methods database is truly a valuable tool for reducing the time and increasing the access to published information on chromatographic and mass spectrometric methods as applied to solving problems with LC/MS. Method developers can access published methods, separation conditions, compounds, compound classes, instrumental techniques, and acquisition parameters. With the click of a mouse, you will be able to search and locate methods from over 13,000 published articles on your:

Compound types and classes	LC/MS options and expected results
Instrumental type and setup	Mass spec acquisition conditions
Separation requirements	Data processing and reduction requirements
Sample preparation requirements	Potential limitations and interferences

First edition, 2006   ISB: 0-9660813-4-x

Order information is provided on the next page.

Order Form

🖷 **FAX Orders:** 1-412-828-3192

☎ **Telephone Order:** 1-412-828-3191
1-888-LCMScom (1-888-526-7266)
(Have your AMEX, Discover, VISA, or MasterCard ready)

🖳 **On-line Orders:** E-mail: books@LCMS.com
Internet: www.LCMS.com

🖃 **Postal Orders:** Global View Publishing
655 William Pitt Way
Pittsburgh, PA 15238

Please Send:
___ Copies of **Mass Spec Desk Reference**, 2nd ed., at $39.95 each
___ Copies of **A Global View of LC/MS**, 2nd ed., at $49.95 each
___ Copies of **LC/MS Methods** at $149.95 each

I understand that I may return any books for a full refund for any reason; no questions asked.

Name: _____

Company Name: _____

Address: _____

City: _____ State: _____ Zip: _____ - _____

Telephone, U.S./Canada: 1- (____)_____

Tele., Foreign: Country Code: _____ No. _____

Shipping & Handling (per book):
USA, Ground   (3–10 days) $ 6.00
USA, Priority   (2–5 days) $12.00
International   (3–10 days) $12.00 (Global Priority or Air)
S&H on multiple book orders is postage plus $6.00 per order.

Payment:
❏ Check (Payable to Global View Publishing)
❏ Credit Card: ❏ VISA ❏ MasterCard ❏ AMEX ❏ Discover

Card Number: _____

Name on Card: _____ Exp. Date: _____/_____